About Sustaining Supply Technology
for Manned Spacecraft

 Projekt Spaceship

ABOUT SUSTAINING SUPPLY TECHNOLOGY FOR MANNED SPACECRAFT

A design concept for extended space travels.
Spaceship classes LiDis I-II.

Christian Zschoch
Projekt Spaceship, 85375 Neufahrn, Germany

Neufahrn, Germany, 2020

Bibliographic information published by the Deutsche Nationalbibliothek:
The Deutsche Nationalbibliothek lists this publication in the Deutsche
Nationalbibliografie; detailed bibliographic data are available on the Internet at
http://dnb.dnb.de.

© 2020 Christian Zschoch
Produced and published by: BoD – Books on Demand GmbH, In de Tarpen 42,
22848 Norderstedt

Printed in Germany
ISBN: 978-3-75284-289-0

Declaration of Authorship

I hereby declare that the thesis submitted

"About Sustaining Supply Technology for Manned Spacecraft"

is my own unaided work. All direct or indirect sources used are acknowledged as references.

I am aware that the thesis in digital form can be examined for the use of unauthorized aid and in order to determine whether the thesis as a whole or parts incorporated in it may be deemed as plagiarism. For the comparison of my work with existing sources I agree that it shall be entered in a database where it shall also remain after examination, to enable comparison with future theses submitted. Further rights of reproduction and usage, however, are not granted here.

Neufahrn, January 10th 2020   _____
city, date   signature

Abstract

At the present time, space travel is characterized by separately developed technologies of the space-traveling nations. Depending on fixed financial budgets and expensive technology companies, the developed spaceships are strongly designed just for a specific mission profile in order to reduce costs and risks as far as possible. Because of their less sustainable supply concept, these spacecraft allow only a limited mission duration and require regular supply deliveries in addition.

On the other hand side, mission periods continue to lengthen with the planned exploration of Mars, asteroids or other objects that are even more distant. These missions will require high sustainable supply concepts in order to enable autonomous and long-term life support of human mission participants. The now existing solutions do not yet meet these requirements, so the current approach of spacecraft design had to undergo a conceptual review.

The research made in the context of this work led to the design of a new generation of spacecraft, which supports with its optimized hull construction such extended long-term missions in terms of durability, variability and life support. All its embedded biological and chemical processes have, on the one hand, the primary aim to enable humans a long stay in space and, on the other hand, to be independent of an external mission supply. The performed research activities also included the necessary mechanical and energetical functions for which an extreme lifetime extension of up to 60 years has been aimed.

In comparison with the already existing designs for large-scale space stations or spacecraft having a low integration of sustainable support systems, the spaceship concept presented here offers a much higher compactness, a lower mass, and a variety of constructively implemented life support functions. The properties of this advantageous spaceship design can also be used stationary on planets, moons or asteroids, if a certain degree of gravity is available after landing.

The scaling of the described spacecraft and its contained systems is in general adjusted to supply one single person. But, however, the therefore required minimum turnovers for plant biomass, drinking water, and nitrogen would also already cover the needs of a second person. Only the preparation of breathing air needs to be enlarged, for which purpose an external source of electrical energy can be added. In order to realize even higher capable spacecraft of this type, the dimensions of the components described can be further increased, which will extend the life support to a higher number of travelers.

Zusammenfassung

In der heutigen Zeit sind Weltraumreisen durch die einzeln entwickelten Technologien der raumfahrenden Nationen geprägt. Gebunden an begrenzte finanzielle Budgets und teure Technologieunternehmen werden die derzeit entwickelten Raumschiffe stark an ein vorgegebenes Missionsprofil gebunden, um die entstehenden Kosten und Risiken weitestgehend zu reduzieren. So erlauben diese Raumflugkörper auf Grund ihrer wenig nachhaltigen Versorgungskonzepte lediglich befristete Missionszeiträume und erfordern zudem regelmäßige Versorgungslieferungen.

Jedoch vergrößern sich die Missionszeiträume mit der geplanten Erforschung von Mars, Asteroiden oder ferneren Objekten stetig, weshalb nachhaltigere Versorgungskonzepte benötigt werden, die eine autarke und langfristige Versorgung von menschlichen Reisenden ermöglichen. Die derzeitigen Lösungen genügen diesem Anspruch nicht, weshalb das aktuelle Raumschiffdesign einer konzeptionellen Überarbeitung bedurfte.

Die im Rahmen dieser Arbeit erfolgten Forschungen führten zum Entwurf einer neuen Raumfahrzeuggeneration, welche mit ihrer optimierten Hüllenkonstruktion solch ausgedehnte Langzeitmissionen in Bezug auf Haltbarkeit, Variabilität und Lebenserhaltung vielfach unterstützt. Alle darin integrierten biologischen und chemischen Prozesse haben das primäre Ziel, dem Menschen einerseits eine lange Aufenthaltsdauer im Weltraum zu ermöglichen und dabei andererseits unabhängig von einer externen Missionsversorgung zu sein. Die ausgeführte Arbeit umfasst auch die hierzu erforderlichen mechanischen und energetischen Funktionen, für welche eine extreme Verlängerung der Betriebsdauer auf bis zu 60 Jahre angestrebt wurde.

Im Vergleich zu den vorhandenen Entwürfen für groß dimensionierte Raumstationen oder Raumflugkörpern mit einer wenig nachhaltigen Systemintegration, bietet das hier vorgestellte Raumschiffkonzept eine höhere Kompaktheit, eine geringere Masse sowie eine vielfach konstruktiv unterstützte Lebenserhaltung. Die vorteilhaften Eigenschaften dieses Raumschiffdesigns sind auch stationär auf Planeten, Monden oder Asteroiden nutzbar, sofern nach der Landung ein gewisser Grad an Gravitation zur Verfügung steht.

Die Skalierung des hier beschriebenen Raumflugkörpers und seiner Einzelsysteme wurde auf die Versorgung einer einzelnen Person abgestimmt. Alle dazu erforderlichen Mindeststoffumsätze zur Nahrungs- und Trinkwasserversorgung sowie des Stickstoffkreislaufes decken jedoch auch bereits den Bedarf einer zweiten Person. Dazu muss für die Atemluftaufbereitung einzig eine externe elektrische Energiequelle ergänzt werden. Zur Realisierung noch leistungsfähigerer Raumflugkörper dieser Bauart kann die Dimensionierung der beschriebenen Komponenten weiter vergrößert werden, um die Versorgung auf weitere Reisende zu erweitern.

Contents

List of Figures 11

List of Tables 13

List of Abbreviations 14

Introduction 16

1 Microbiology as a Driver for an Organically Integrated Spaceship Concept 17

2 Outer and Inner Structure of a Self-Sufficient Spacecraft 19
   2.1 A basic Hull Design of a Gravity Supporting Spacecraft 19
   2.2 Inner Spaceship Structures for Functional Purposes 24
   2.3 Hull Wall Construction 28
      2.3.1 Thermal Hull Compensation Flow 28
      2.3.2 Inner Spaceship Walls and Floors 29
      2.3.3 Planking Materials and Weight of the Spaceship Hull 32
      2.3.4 Implementation of Hull Leadthroughs 32
      2.3.5 Meteorite Protection Fabric 33
      2.3.6 Thermal Hull Insulation 34
      2.3.7 Radiation Protection Measures 37
      2.3.8 Qualitative Hull Checks 40

3 Systems for Life Support 41
   3.1 Plant Species for Space Planting 41
   3.2 Hydroponic Planting 43
   3.3 Nutrient Solution Management 45
   3.4 Nutrient Solution Circulation 51
   3.5 Illumination of the Plantings 54
   3.6 Photosynthetic Respiration Air Regeneration 57
   3.7 Respiration Gas Pressure Changes 59
   3.8 Human Nutrition and Planting Management 62
   3.9 Salt Extraction 65
   3.10 Drinking Water Condensation 66
   3.11 Air Circulation Management 68

4 Systems for Energy Supply and Organic Material Processing 70
   4.1 Internal Energy Supply 70
      4.1.1 Bioreactor Units to the Degenerative Fermentation 71
      4.1.2 Biogas Storage Tanks 75
      4.1.3 Methane Gas Reformer Units 77
      4.1.4 Overflow Lifter Pump 82
      4.1.5 Simplified Hydrogen Fuel Cells 83

4.2 External High Energy Sources..86
    4.2.1 Solar-based Energy Generation..87
    4.2.2 Radioisotope Generator...88
    4.2.3 Cold Fusion Reactor..91
4.3 External Material Collection..97
    4.3.1 Hydrodynamic Material Lock..99
    4.3.2 Static Charge Generator..101
4.4 Energy Dispatching Nets..101

5 Simplified Apparatuses for Electro Mechanics, Navigation and Spacecraft Propulsion..102
5.1 Internal Drives, Environment Sensors and Navigation Solutions..........................102
    5.1.1 Hydromagnetic Bearings, Floating Direct Current Motors and Magnetic Gearwheels...102
    5.1.2 Navigation Projection System..105
    5.1.3 Environment Sensors and Control Instruments...108
    5.1.4 Optical Display Scanning and Rotation Balancing....................................111
    5.1.5 Antennas and Radio Communication System...115
5.2 Spaceship Propulsion...116
    5.2.1 Space Engine Suspension..116
    5.2.2 Review of alternative existing Propulsion Solutions.................................118
    5.2.3 Centrifugal Mass Space Engine...119
5.3 Personal Living Space...124

6 On Board Software...125
6.1 Electronics Hardware...125
6.2 Computer Operating System..125

7 Conclusions..132

Acknowledgements...133

Bibliography...134

List of Figures

Fig. 1a: Top view of the outer hull structure..22
Fig. 1b: Side view of the outer hull structure..22
Fig. 1c: Rear view of the outer hull structure..22
Fig. 1d: Single frame profile with size ratio..23
Fig. 2: Mounting of the landing gear..24
Fig. 3a: Top view of the inner structure..27
Fig. 3b: Side view of the inner structure..27
Fig. 3c: Rear view of the inner structure and the MCB..28
Fig. 4: Circulation of the thermal compensation flow through a spaceship segment........29
Fig. 5: Insulation and radiation shielding layers of the spaceship hull..36
Fig. 6: Pivoting mechanism of the planting racks..43
Fig. 7: Cross section of a hydroponic planting channel..45
Fig. 8: Rate of ammonium to nitrate conversion and following denitrification...............47
Fig. 9: Cut view of a nutrient solution transport channel for variable gravity..................51
Fig. 10: Archimedean screw pump for pumping the nutrient solution..53
Fig. 11: Diagram to the growth rate, light stick distance and illumination intensity........55
Fig. 12: Influence of the ambient air by the oxygen, carbon dioxide and nitrogen cycle..61
Fig. 13: Cut view of two water condensers with an underlying bioreactor-fuel-cell-unit. 67
Fig. 14: Cut view of ventilation unit with floating direct current motor..........................69
Fig. 15: Air flows in the concept spaceship..70
Fig. 16: Arrangement of the bioreactor-fuel-cell-units..72
Fig. 17: Cut view of a bioreactor..73
Fig. 18: Cut view of a biogas tank..76
Fig. 19: Counterflow reformer with inlet funnel for hydrogen production......................80
Fig. 20: Cut view of the overflow lifter pump..83
Fig. 21: Single cell of a fuel cell block..84
Fig. 22: Integration of the radioisotope generators..90
Fig. 23: Principle of the electrostatic nuclear fusion..95
Fig. 24: Principle of an electrostatic acid fusion..95
Fig. 25: View of the MCB..98
Fig. 26: Hydrodynamic material lock..100
Fig. 27: Cut view of the floating direct current motor..102
Fig. 28: Circuit of the floating direct current motor..103
Fig. 29: Cut view of the nutrient solution pump bearing..104
Fig. 30: Functional scheme of a magnetic gear..105
Fig. 31: Navigation projections on the CINA..106

Fig. 32: Arrangement of the navigation room..107
Fig. 33: Arrangement of the CINA instruments..111
Fig. 34: Sensor arrangement for movement detection..113
Fig. 35: Denitrification and balancing tank system...114
Fig. 36: TABAS/LCP computer simulation..115
Fig. 37: Transmitting- and receiving antennas at the spaceship front............................116
Fig. 38: Side view of a gravitation pendulum...121
Fig. 39: Gravitation pendulum endpoints..122
Fig. 40: Principle of a mass-centrifugal engine...124
Fig. 41: Screenshot of a CLEO application..127
Fig. 42: CLEO thinking process..128

List of Tables

Table 1a: Material list outer frame 30
Table 1b: Material list inner structure 31
Table 1c: Material list outer planking 31
Table 1d: Material list inner planking 31
Table 2: Light intensities of the planting lights, depending on the growth height 57
Table 3: Plant density and biomass portions 58
Table 4: Oxygen and carbon dioxide balance 59
Table 5: Daily calorie requirement of a person within the concept spaceship
(BMR = 49 kcal/h) 62
Table 6: Daily nutrient supply of a space traveler 64
Table 7: Thermal energy balance 81
Table 8: Internal electrical energy balance 85
Table 9a: Sensors and controls for flight operations 109
Table 9b: Sensors and controls for the environment 110
Table 9c: Sensors and controls of the power supply 110
Table 10: Sensor detection of celestial objects 112
Table 11: Composition of the spaceship mass 120
Table 12: Energy demands of the CLEO system 126
Table 13a: Threshold variables of the CLEO kernel 129
Table 13b: Example parameter of a logic word 130
Table 13c: Logic connections of single logic table entries 131

List of Abbreviations

General abbreviations:

LiDis n	Living Disk
	Spaceship class, specifying the maximum number of supported persons
LED	Light Emitting Diode
LCD	Liquid Crystal Display
ARM	Advanced RISC Machines

Spaceship components:

OHS	Outer Hull Structure
CT	Central Tunnel
LT	Lengthwise Tunnel
MCB	Material Collection Bay
NAV	Navigation Room
BFT 1-2	Biological Fermentation Tanks
GAT 1-2	Biogas Tanks
CWT 1-4	Clean Water Tanks
DBT 1-4	Denitrification and Balancing Tanks
PU 1-4	Propulsion Units
OS 1-8	Outer Segments
	Outer spaceship segments in clockwise order, starting with the first segment it front direction
IS 1-8	Inner Segments
	Inner spaceship segments in clockwise order, starting with the first segment in front direction

Spaceship circuits:

HTC	Hull Temperature Circulation
ISA	Inner System of Air Circulation
PS 1-4	Propulsion Steam 1-4
IBE	Internal Biological Energy
RBE	Radioactive Decay Based Energy
FBE	Fusion Based Energy

List of Abbreviations

Functional units:

CINA	Central Installation for Navigation and Automated Processes
CLEO	Computers Local Environment Operator
TABAS	Tank Balancing Adjustment System
LCP	Light Controlled Propulsion
GSC	Generator for Static Charge
MRU 1-2	Methane Reformer Units
SFC 1-2	Submerged Fuel Cells
WCU 1-4	Water Condensing Units
NSPU	Nutrient Solution Pump Unit
DCM 1-2	Floating Direct Current Motor
HCF	Hull Circulation Fan
ICF	Inner Air Circulation Fan

Introduction

Space flight raises a great fascination on us humans since its origin. And even in the days before, futurologists and scientists of many countries were planning and describing journeys to far distant planets, asteroids, or solar systems. At all these times, such thoughts were influenced by the available technical possibilities, whereby all developed exploration plans had to be rejected soon or later.

Only today, for the first time in history, the available technologies are so advanced that enlarged space travels seem to be possible. Having now this technical opportunities, mission objectives like a manned permanent lunar base, asteroid visits, or a flight to Mars are coming more and more into the reach of mankind. These upcoming long space flights will require various innovative and sustainable spacecraft system solutions for a large number of supply issues, which then have to be integrated into a universal spaceship design. All these individual solutions must be properly dimensioned and be balanced to each other, so that – depending on the mission duration, the mission target, and the number of space travelers – closed material and energy cycles are created. Moreover, require such manned long-term missions a higher consideration of human kind of living and the physiological needs of all embedded organisms to survive such journeys as healthy as possible.

Due to the lack of super-fast space engines, whose technology is unlikely to become real in the foreseeable future, the idea of generation ships, as they were conceived by the physicist John Desmond Bernal in the 1920s, are probably the most technically plausible method to make a journey to destinations outside our solar system (Bernal *et al.* 1929). Bernal's vision of a multi-hundred-meter-long, rotating spaceship cylinder, which could be used as a permanent home for tens of thousands of people, embodies already the basic idea of the integration of sustainable life support components into a permanently used spacecraft.

The design concept presented here, also takes up this approach and combines many miniaturized and simplified spaceship components, making the spacecraft based on it smaller and – with a reduced crew – also cheaper to build and maintenance.

During the development of this design concept, varied research activities have been executed since 2007, out of which the results are summarized in this document. The experiments for building up a data basis to the different topics of spaceship design, support the made assumptions with verifiable results and thus could be used for a detailed simulation of the individual systems. As a result out of this experimental design phase, also already existing solutions of current spaceflight have been discarded, or more suitable alternative solutions have been found for them.

Even only the number of topics considered for this work is both overwhelming and intriguing. From bio-chemical and physical processes, to the cultivation and degeneration of plants, from optical lens systems to the programming of processor cores, the palette reaches. In order to keep an overview of all activities and to structure the development work, the different design topics have been divided into diverse sub-projects. Their related detail documentation was recorded in a structured form and was generally made available via an internet homepage.

These results can be viewed on the website WWW.PROJEKT-SPACESHIP.DE.

The design of the concept spaceship described in this document is protected by copyright laws with all its individual developed components. In order to grant a permission for a commercial use, please ask the author directly or use the related form of the project homepage.

1 Microbiology as a Driver for an Organically Integrated Spaceship Concept

If a permanent supply for space travelers has to be implemented into a completely closed spaceship environment, the holistic consideration of all organic interactions within the surrounding ecological system has to be discovered first. This concerns the microbiological processes of food growing and processing as well as the respiratory influence on the oxygen and carbon dioxide content in the air.

The life support systems on board a long-term used spaceship must support all possible human material cycles in order to prevent a shortage of life essential raw materials. The main material cycle begins with the decomposition of excreta and waste in bioreactors, which disintegrates the introduced substrate via anaerobic digestion and produces a usable amount of methane gas at the same time. Analogous to the natural recycling processes, the remaining and liquefied residues can subsequently serve for the nitrification of a circulating nutrient fluid, which can be used for a hydroponic plant cultivation. While the so inserted plants are growing, they continuously regulate the levels of carbon dioxide CO_2 and oxygen O_2 in the respiration air, but they also have their own demands on their environment, which is why additional components for energy, lighting, ventilation, and heating are required. And last but not least, it is moreover important to involve a person, who continuously drives all these processes and supports them with their microbiological material processing.

If the biological process chains are not completely integrated into a sustained life support system, and if one of the material cycles ends at a certain point, the system will automatically reach a shortage situation after a limited system runtime, since required basic substances cannot be provided to the following processes at this place. And because most of these processes are based on microorganisms, the failure or absence of even one single integrated life form can cause an artificially material balance to tip over.

An example of such a balance disturbance is known from the Biosphere 2 project, which was used to study a capsuled closed ecosystem and was in 1994 successful in a hermetically sealed operation for a period of six months. Some months before, a first long-term trial failed during the years 1991 to 1993, as the concrete walls in the system performed an unexpected assimilation of oxygen, which made the experiment dependent on an external oxygen supply. Irrespective of this, the project impressively showed how a large number of organisms could take over the full supply of eight persons, which was very close to the ideal picture of a closed material loop (Nelson *et al.* 1994).

In contrast to this, there are also examples that argue against excessive incorporation of life forms into a life support system. Because if an organism does not find ideal living conditions in a closed environment, evolutionary and unpredictable changes in the metabolism of these life forms will modify their behavior. A well-known example is thereto the uncontrolled colonization of the Mir space station with mutated microorganisms that had the ability to corrode aluminum and plastics of the station construction (Novikova 2004). In such a situation, not only the physical health of the astronauts is in danger by the excessive exposure to microbe-enriched air, it is also very difficult to keep the spacecraft structurally intact, as it was realized in the case of the Mir space station after only 15 years of use.

Because the material research provides only partially corrosion-resistant materials for such a case, material thicknesses with sufficient reserves have to be planned for a spaceship construction in order to allow a maximum mission duration that corresponds to four times the service life of the Mir space station. In addition to this, can electrical and electromechanical systems be encapsulated in plastics or integrated in a simplified version, so that massive electrical insulations and circuit-short proof distances between electrical conductors are ensured. With the usage of such resistant solutions, as they are mostly described below, the operation of the foreseen spacecraft systems can be permanently ensured even in a microbiological high active environment.

As a conclusion of the microorganismic analysis, it can be also stated that in the design of a sustainable life support system in general a variety of specific microorganisms is required to manage all metabolic processes, but only the absolutely necessary organisms should be considered for this, which have a controllable growth rate and will remain in their defined biological habitat. This reduction of biodiversity – together with the corresponding reduction of necessary habitat types – enables the radical simplification of the Biosphere 2 approach: a noticeable reduction of volume and weight for a sustainable life support system, since a large-scale solution would be far away from the claim of this work to design a compact, self-sufficient spacecraft.

2   Outer and Inner Structure of a Self-Sufficient Spacecraft

The requirement for an ideally compact and lightweight spaceship construction was at all times challenged during its design process by the demands of its passengers, who were the largest living organisms to be considered. Therefore, with the design of the here introduced spaceship hull the basic requirements were created for

- the use of an existing gravity as well as the generation of an artificial gravity,

- sufficient space to implement a planting and life support system,

- the optional installation of an airlock,

- an optimal positioning of the navigation- and communication systems, as well as for

- a multi-layered spaceship hull structure supporting a temperature management in case of one-sided solar heat exposures, general thermal insulation, radiation protection and prevention from meteorite impacts.

2.1   A basic Hull Design of a Gravity Supporting Spacecraft

In its main function, the outer hull structure (OHS) provides a solid construction for the integration of all components and attachments. The basis for this will be a support structure made of aluminum profiles, which run along all spaceship edges. The degree of hardness and the material thickness of the implemented beams should be chosen so that the construction is as lightweight and flexible as possible. For the design concept, a square frame profile with a cross section measurement of 10 x 10 centimeter and a material thickness of three millimeters has been assumed. The connection of the profiles will be carried out by welding whereby an optimal frame strength is reached.

In order to create an optimized spaceship body for human usage, specific minimum sizes and a special design are advantageous. The basic shape is formed by a circle that can be easily constructed out of curved frame components. Along its circumference, this shape has the properties of a gyro, which are very advantageous for the later operation. On Earth or in a generally landed state, the round circle disk serves as a walking floor. Out of this flat surface, a disk-shaped body has been constructed by putting two of those surfaces in parallel to each other and connecting them with 16 vertical struts along their outer circumference. Thereby, the distance between the two circle surfaces is ideally 2.3 meters, which allows for humans an upright walking. The vertical, outer struts are placed in a constant distance to each other and are made by an alternate use of solid 10 x 10 centimeter profiles and 5 x 10 centimeter double-T profiles with central openings. The combination of these profiles saves on the one hand weight, but it also allows a controlled thermal circulation through the inserted openings within the outer hull.

During a flight in space – without the influence of an external gravity – the disk-shaped design now has a decisive advantage: a spaceship of this type can be rotated around its vertical center axis without changing any hull configuration, creating a centrifugal force along its circular side surfaces. By providing this force, this simple spaceship design provides a very easy way to generate an artificial gravity that allows an upright movement for humans on the inner side surfaces.

The diameter of the OHS disk should be dimensioned so that a human upright walking is also possible during a rotation phase of the spaceship. Further, the circular side surfaces should be as far away from the center of rotation as possible so that the required rotation speed can be lower. If we assume an internal rotated corridor height of 2.60 meters, the spaceship would only have a diameter of just over five meters. In order to increase the distance between the outer surfaces and the rotation center in addition, a second circular inner corridor was added to the construction, having a radius respectively a rotating corridor height of 2.07 meters. Out of this results that the rotating spaceship consequently provides an outer and inner circular walking floor. If the thickness of the required insulation, which requires in average 25 centimeters at all sides, is considered as well, the spaceship diameter of the outer hull structure will reach a total of 9.81 meters.

In order to generate an artificial gravity of 80 percent of the Earth's gravity along the outer circumference surface, a centrifugal force corresponding with 80 percent of the gravitational acceleration is needed. Measured by a kilogram of mass, this will mean a weight force of 7.848 Newton. The formula of centrifugal force is first resolved according to the required circumference velocity v and then calculated with the given Earth acceleration and spaceship radius:

$$v = \sqrt{\frac{F \cdot r}{m}} \, m/s$$

$$v = \sqrt{\frac{7,84 \, N \cdot 4,905 \, m}{1 \, kg}} \, m/s$$

$$v = 6,2 \, m/s$$

With this value, the needed number of revolutions then can be determined by dividing the spaceship circumference with it and by converting the result into revolutions per minute.

$$n = 60\ s \div \frac{2 \cdot \pi \cdot r}{v} RpM$$

$$n = 60\ s \div \frac{2 \cdot \pi \cdot 4.905\ m}{6.2 m/s} RpM$$

$$n = 12.07\ RpM$$

Thus, to provide the intended artificial gravity along the outer hull surface, twelve revolutions per minute are required.

Due to stability reasons, the upper and lower deck surface of the spaceship disk cannot be realized as flat structure, because the inner air pressure pushes very strongly against these surfaces, and the hull material can therefore easily be overloaded. The hereto chosen solution is the usage of domed deck surfaces, which – similar to the ends of a cylindrical gas tank – absorb the forces of the internal pressure evenly. To construct the deck surfaces the spaceship rotation axis was elongated at both ends by 50 centimeters and was connected with 16 curved horizontal struts to the ends of the 16 vertical struts of the circumference surface. Analogous to the changing strut profile of the outer side wall, also this deck structure has been implemented at every second strut with the use of an opened double-T profile, while all other struts consist of the 10 x 10 centimeter profile.

In order to reduce the air resistance during atmospheric flights, but also to provide additional space for the required communication antennas, a front-pyramid made of further frame profiles was added to one side of the spaceship body, which extends over a range of four vertical side wall struts. The top of this pyramid is located two meters above the regular outer wall surface in a centered position. Its four pyramid edges are also made of 10 x 10 centimeter frame-profiles, while all inner struts are made of flat material, which creates less interference with the radio signals of the communication antennas underneath.

Starting from the spaceship front-pyramid the spaceship disk is divided into eight segments, which are – viewed from above – clockwise numbered in ascending order. Each segment is thereby defined by the space between two 10 x 10 centimeter outer frames and extends consequently over two vertical outer struts.

After this numbering, the segments can be assigned to the following functions: the segment 1, pointing to the spaceship front, is a residential and control segment. The two right and left adjoining segments 2 and 8 are planting segments, for which each the attachment an external engine gondola has been foreseen. These segments are followed by the two middle segments 3 and 7 – both pure planting segments – and the segments 4 and 6, which were again designed as combined planting and engine gondola segments. The rear segment 5 will be used as a connection segment for different tasks.

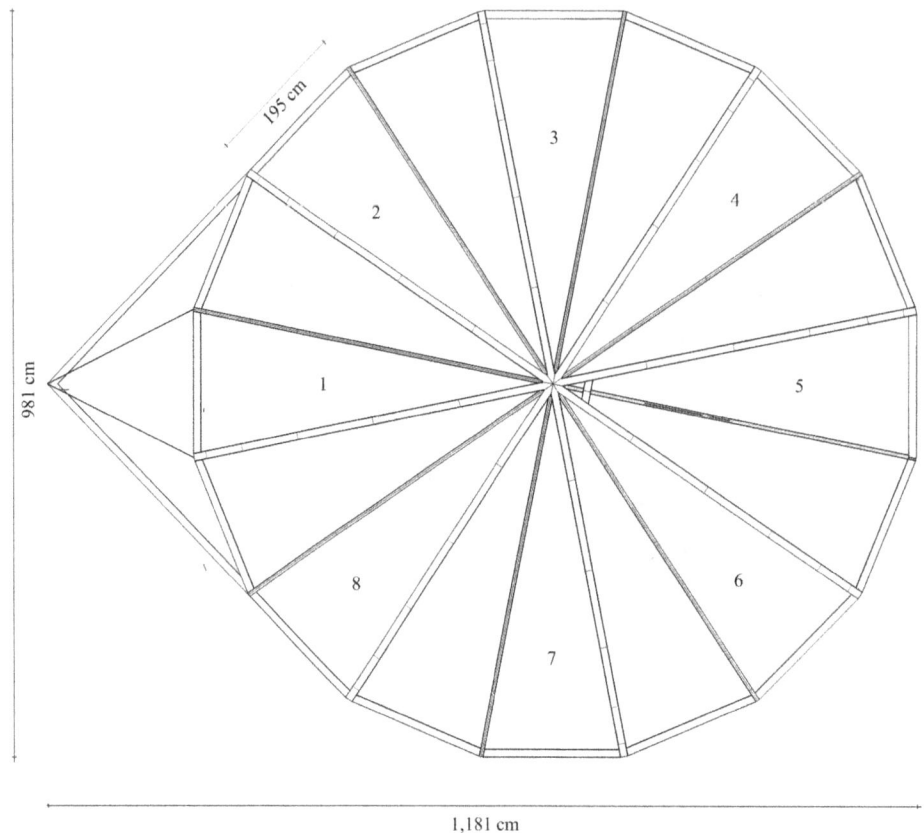

Fig. 1a: Top view of the outer hull structure

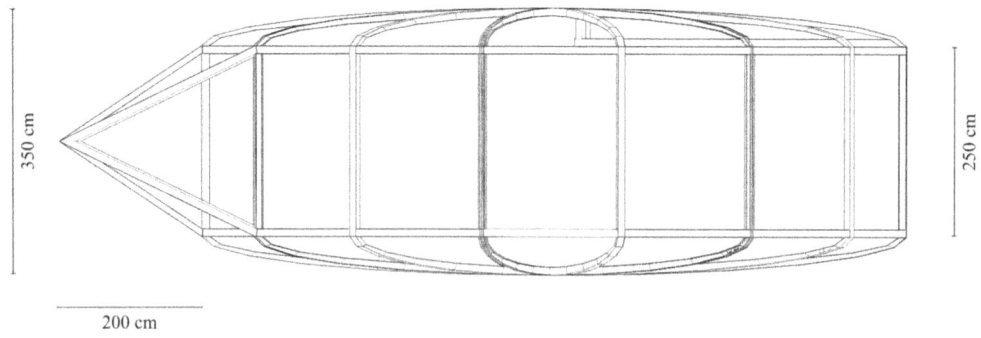

Fig. 1b: Side view of the outer hull structure

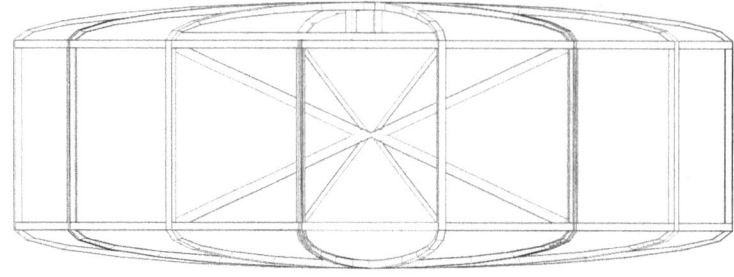

Fig. 1c: Rear view of the outer hull structure

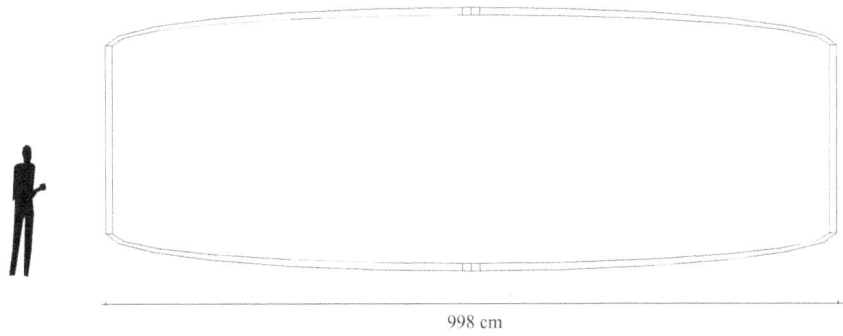

Fig. 1d: Single frame profile with size ratio

By the constructional and functional assignment of the spaceship segments, a balanced mass distribution was reached around the central spaceship axis. This is necessary in order to minimize the imbalance to be compensated during a rotation flight. For example, the six planting segments are distributed to both sides of the central axis with each three segments, and also the additional weight of the spaceship front-pyramid will be compensated in the following by a special inner structure in segment 5. All other internal and external components, like bioreactors, water tanks, or engine gondolas, were arranged according to this mass distribution principle as well in order to minimize their influence on the rotational balance.

In landing configuration, the stable stand will be ensured by three retractable landing gears, which are mounted in the lower fore deck surface and at the central vertical struts of the segments 4 and 6. The front landing gear is centered on the lengthwise axis of the spaceship and is placed directly in front of the planned ring corridor separation wall of the outer segment 1. It is connected to the outer frame construction by two vertical and two horizontal double-T beams. The outer hull wall inclusive its insulating layers is here indented to the inside and surrounds the landing gear up to these inner mounting struts so that an air-sealed and insulated landing gear housing is created.

The two rear landing gear housings fit with their size in the space of the hull insulation. These landing gears use additional profiles as well to connect them firmly to the spaceship frame.

The front landing gear housing extends 1.06 meters, the rear landing gear housing 0.82 meters into the hull inside, wherein the front housing has a cross-section of 30 x 30 centimeters and the two rear housings of 18 x 18 centimeters each. For a better mobility, the mechanics of the front landing gear should be provided with a steering function. Furthermore, the landing gear construction must take into consideration the expected weight load, which is determined during the following sections.

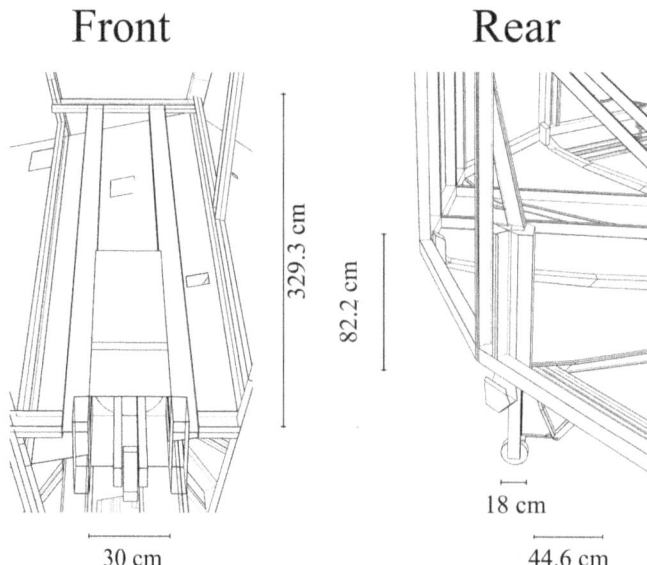

Fig. 2: Mounting of the landing gear

The concept described here intends the usage of four space engines, which can be operated though with a low power only, but for a very long period of time. The permanently generated thrust of these engines shall enable just a slow lifting of the spaceship and a slight acceleration against the Earth gravity. Each of the four engines is – in dependence to the chosen type of drive – encapsulated when possible in an engine gondola housing and is mounted at the outer side of the spaceship. For the engine gondola mounting, large rotatable ring bearings are required in the outer side wall of the segments 2, 4, 6 and 8, which are welded into these surfaces and connected to the frame construction via struts in addition. The rotatability of the engine gondolas mountings is necessary, since they either shall be used in a horizontal orientation to influence the gravity rotation, or in a vertical orientation to perform the propulsion of the spaceship disk along its rotation axis.

2.2 Inner Spaceship Structures for Functional Purposes

Inside the outer hull structure, additional struts are inserted for an internal structure (IS) that further increases the stiffness of the hull construction and spatially separates the areas of use. The material used for the inner structure is again an aluminum material in form of the already used double-T profiles with openings and a material thickness of three millimeters.

The so-called central tunnel (CT) has been inserted first. It extends around the axis of rotation and has a diameter of 80 centimeters. For its construction, eight double-T profiles are welded vertically between the outer 10 x 10 centimeter frame profiles. All struts point with their narrow side towards the tunnel center for maximizing the stability of the tunnel wall applied thereon.

In order to use the central tunnel for the intended optical navigation, massive glass lenses are inserted at its upper and lower end through the middle of the in star shape connected outer frames, thereby enabling a followed projection of the incoming starlight. Particularly important are the projections out of the central tunnel axis during the different flight phases, when a target object has to be aligned along the rotation axis direction, or if the rotation balance of the spaceship needs to be monitored continuously. For the final projection, the incoming light is further bundled by a lens combination within the tunnel and deflected by two mirrors into the navigation room, where a projection screen and a sensor field are located. Further information about the navigation technology as well as the possible engine types are further described in the chapter on engines and navigation.

Another advantage of the central navigation tunnel is the avoidance of spaceship areas that have to be excluded from the gravitational rotation. In many existing concepts for rotating spaceships such areas are foreseen as engine- and fuel tank unit, or they are used for control- and navigation purposes. For example, the NASA concept of the Nautilus-X spaceship plans to use a non-rotating central spaceship body with weightless areas, while a residential unit rotates around this to generate an artificial gravity (Holderman and Henderson 2011). However, in order to allow the separately rotated unit a free movement, this spaceship design will require additional bearings, hatches and seals, which during a long-term mission underlie a corresponding aging as well as a continuous abrasion and thereby increase the risk of failures.

In contrast thereto, the here described spaceship concept offers a combined gravity, navigation and propulsion solution, in which one single and compact spaceship body provides all these functionalities and thus minimizes the mentioned failure risks as well as the necessary maintenance for the spaceship construction.

Another characteristic feature of the outer hull structure is located at the top side of segment 5. Starting from the outer edge of the spaceship, a flat surface extends here to the wall of the central tunnel. The edges of this surface are supported by two inner horizontal 10 x 10 centimeter profiles, which are welded into the outer frame and that follow the horizontal surface from the outer side wall to the central tunnel struts. The center of the surface is additionally strengthened by a dividing middle strut that is inserted therein as replacement for the missing upper curved double-T beam. Because this indentation is a part of the outer wall, it will be fully planked and insulated with the selected hull materials as well. The space therein is the so-called material collection bay (MCB) that during rotation flight can collect cosmic dust from space by using a static charge. For this purpose, the MCB has a moveable one-side-opening cover, which follows in its closed position the curved hull shape, and which in opened state lifts a static charged and finned panel on its inside. The details of the MCB are described further in the chapter on energy supply and organic matter processing.

Below the MCB a longitudinal tunnel (LT) is located, which extends along the longitudinal axis of the spaceship through the entire segment 5 and thus reaches from the central tunnel to the rear vertical outer side wall. In order to integrate this structure, two double-T profiles were inserted vertically into the frame of the outer side wall of the 5th segment. The distance between these profiles, which point with their narrow sides in the direction of the spaceship longitudinal axis, is 79.2 centimeters, whereby the middle of the so created opening is located exactly on the spaceship longitudinal axis. Starting from the upper and lower ends of these profiles, four parallel aligned, horizontal double-T struts has been inserted, which point with their narrow side upwards and with their inner end are welded to the central tunnel. The walls of the longitudinal tunnel are each strengthened by a vertical double-T strut at the position of the ring corridor separation wall as well as by two further diagonal struts, and they will be planked on their inner side by a single layer wall coverage.

The longitudinal tunnel serves in the concept spaceship for several purposes. First of all, its construction forms a counterweight to the front-pyramid and it's transmitting and receiving antennas, which is important for a balanced rotation of the spaceship body. Another main function is its use as an integrated stairwell that connects the inner and the outer ring corridor. For this purpose, the tunnel side walls at each ring corridor level were equipped with a hatch to each side, which ensure a quick access to all spaceship areas.

In a landed state, a traveler can move very easy between the ring corridors by walking on the lower deck surface. For a usage of the longitudinal tunnel during a rotation flight, aluminum ladders are mounted at both walls of the tunnel, which allow an ascent or descent between the two ring corridors under the conditions of artificial gravity as well. And because all openings of this tunnel can be equipped with airtight hatches, it can serve in its last function also as an airlock or docking gate, but this usage of course depends on the available amount of compressed air for the necessary restoring of the inner pressure.

The internal structure also includes the physical separation of the two ring corridors around the center of rotation. For this, eight vertical double-T profiles were inserted, which will be welded between the outer upper and lower 10 x 10 centimeter frame profiles, having a distance of 2.56 meters to the vertical outer struts. For reasons of stability, their narrow sides point to the spaceship center. In order to integrate a stable ring corridor wall of non-curved surfaces, eight further vertical struts have been added into the outer frame double-T profiles in-between, having a slightly increased distance of 273.2 centimeters to the outer struts.

A last component of the inner structure are eight 1.5-meter-long planting rack holders, which connect diagonally, in every segment-limiting outer frame, the vertical strut of the ring corridor separation wall with the upper segment frame profile. The used double-T beams are thereto inserted with their narrow side upwards and 45 degrees rising inclination in the outer ring corridor, so that the hull construction receives some additional stiffness, too. In the planting segments, these struts are provided on their upper side with centered axis bearings whereby a flexible mounting point for the pivoting plant racks is formed.

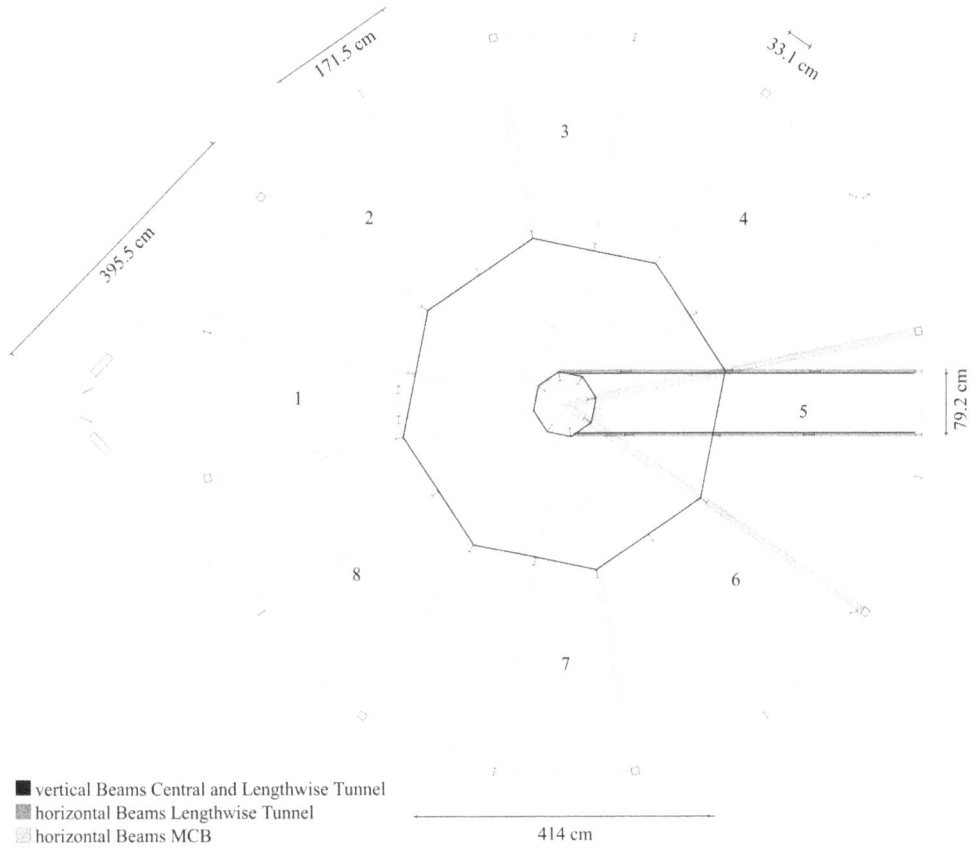

Fig. 3a: Top view of the inner structure

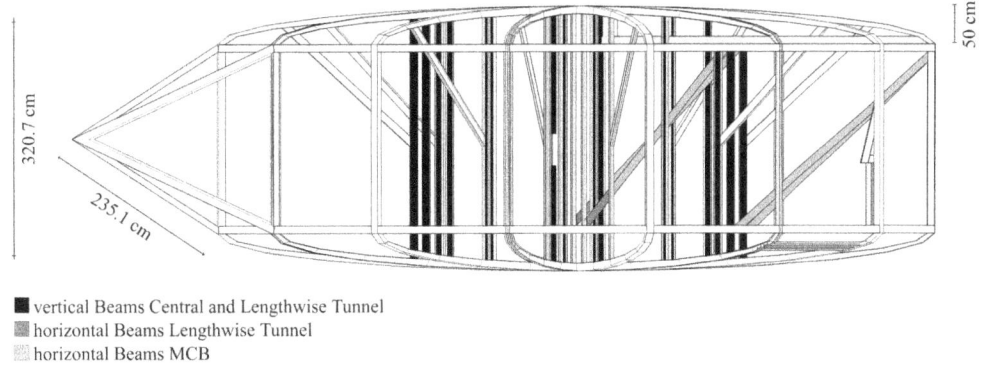

Fig. 3b: Side view of the inner structure

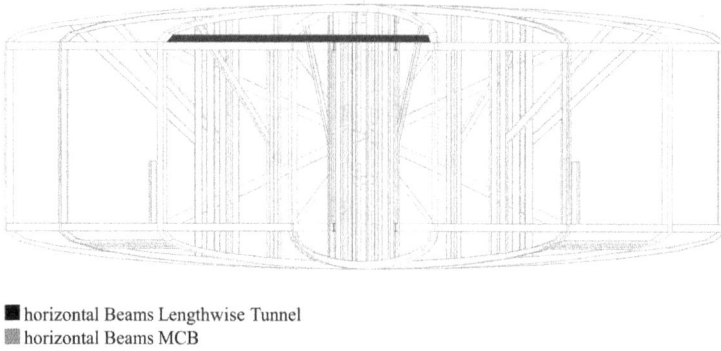

■ horizontal Beams Lengthwise Tunnel
▨ horizontal Beams MCB

Fig. 3c: Rear view of the inner structure and the MCB

2.3 Hull Wall Construction

The frame construction is externally planked by aluminum sheets, having a material thickness of ten millimeters. All planking fugues are carried out pressure-tightly, so that an airtight spaceship hull is created that is only imperfected by a central hatch in segment 5 and a few supply and lens leadthroughs. Other openings, like additional windows or hatches, are not foreseen in order to minimize the risk of a pressure loss caused by ageing or damage.

In parallel to the outer hull planking, the spaceship body, including the surfaces of the MCB, has also an inner covering with a material thickness of three millimeter, which is connected to the inner sides of the outer frame construction. This inner layer will be in the following completed by the adjoining insulation and radiation protection layers.

2.3.1 Thermal Hull Compensation Flow

In addition to the increased hull stiffness, this double planking also forms a hull cavity that allows the implementation of an inside all-around air or water circulation, which compensates temperature fluctuations of the spaceship hull along its outer surfaces. Inside a heliocentric solar system, the sunlight hits a spaceship mostly from one side, causing very high temperature differences in its hull (Birur *et al.* 2001). The thereof resulting different expansion of the outer hull materials would lead to strong material stresses in the construction, which should be avoided. The implemented thermal compensation flow, that is operating in parallel to the passive insulation and reflection layers, makes it possible to balance such temperature differences in order to ensure a mostly homogeneous hull temperature.

For a continuous connection of the adjoining hull-segment cavities, pipe-connectors are mounted at the upper and lower inner-edge of each segment that direct the air or water flow – coming from the upper deck cavity – first into a corner of the outer side wall cavity. From there, the thermal flow extends diagonally through the side wall and reaches

Table 1b: *Material list inner structure*

Double-T profile 5 x 10 centimeter with openings:	
8 x Central tunnel strut, vertical à	3.29 Meters
4 x Longitudinal tunnel strut, horizontal à	4.81 Meters
2 x Longitudinal tunnel strut crosswise outer segment à	3.40 Meters
2 x Longitudinal tunnel strut crosswise inner segment à	3.23 Meters
2 x Rear landing gear suspension complete à	5.34 Meters
1 x Front landing gear suspension complete	8.25 Meters
8 x Planting rack holders à	2.36 Meters
Total:	96.63 Meters

Table 1c: *Material list outer planking*

Material for outer planking, ten millimeters thick:	
4 x Inner front pyramid strut 2,55 meters à	0.25 sqm
30 x Hull planking horizontal, curved à	4.92 sqm
16 x Hull planking, vertical à	4.88 sqm
1 x MCB cover, horizontal à	9.84 sqm
1 x MCB outer planking, horizontal à	7.20 sqm
2 x MCB side planking, vertical à	1.20 sqm
Total:	246.12 sqm

Table 1d: *Material list inner planking*

Material for inner planking, three millimeters thick:	
8 x Ring corridor separation wall, vertical à	5.50 sqm
8 x Central tunnel wall, vertical à	1.09 sqm
2 x Longitudinal tunnel wall, vertical à	12.94 sqm
30 x Hull planking horizontal, curved à	4.66 sqm
1 x MCB outer planking, horizontal à	7.20 sqm
2 x MCB side planking, vertical à	1.20 sqm
16 x Hull planking, vertical à	4.78 sqm
2 x Inner wall outer segment 1 à	7.29 sqm
2 x Inner wall inner segment 1 à	5.65 sqm
Total:	330.36 sqm

2.3.3 Planking Materials and Weight of the Spaceship Hull

For a rough weight calculation of the spaceship body, initially the strut and panel surface can be multiplied by the used material thicknesses, which will result in a material volume of about 3.79 cubic meters. In combination with the material density of aluminum of 2,700 kilograms per cubic meter, a rounded hull mass of 10,234 kilograms can so be determined. This specification refers to the basic construction of the spaceship hull and the additional internal structures, but does not take into account either any further radiation protection measures, thermal insulation, internal or external components, organic material, or any needed water.

The usage of aluminum for the planking has been mainly preferred due to its radiation shielding properties. Other construction material alternatives, like carbon fiber plastic, might be able to reduce the construction mass, but then the radiation protection layer has to be strengthened, what again would lead to a mass increase.

Another option for optimizing the spaceship hull weight would be the use of a balloon-like outer wall with only a few additional stiffening structures. Such a hull structure has been already developed for the TRANSHAB concept (De la Fuente *et al.* 2000) of NASA, which was successfully tested with several test modules in space. Furthermore, the cylindrical basic shape of a TRANSHAB module is similar to the disk-shaped form of the spaceship concept described here. If the mass of a TRANSHAB wall – which consists more than 60 layers of different materials – is assumed with a thickness of 19 millimeters and a material density of 1,800 kilograms per cubic meter, the mass of the spaceship hull would be only 8,414 kilograms due to no longer necessary supporting structures.

Because a spaceship hull used for the intended long duration missions should provide a reliable high stability and flexibility, the variant of a full aluminum spaceship hull has been chosen for the further described construction. This will certainly require more powerful engines due to a higher mass, but it uses a reliable construction technique, whose material has already been proven as very durable in space.

However, depending on the necessary spaceship lifetime or the planned target location, also the choice of one of the other described material combinations may be more advantageous.

2.3.4 Implementation of Hull Leadthroughs

Along the hull surface, various leadthroughs are necessary for control and supply purposes, which must be pressure-tightly connected to the outer wall. In order to avoid material fatigue at these leadthroughs, stiff and massive metal lines will be embedded the lower deck cavity at their opposite corner. Since the inner double-T profiles of the segment walls are provided with openings over their whole length, the diagonal flow can pass directly over the entire width of a segment.

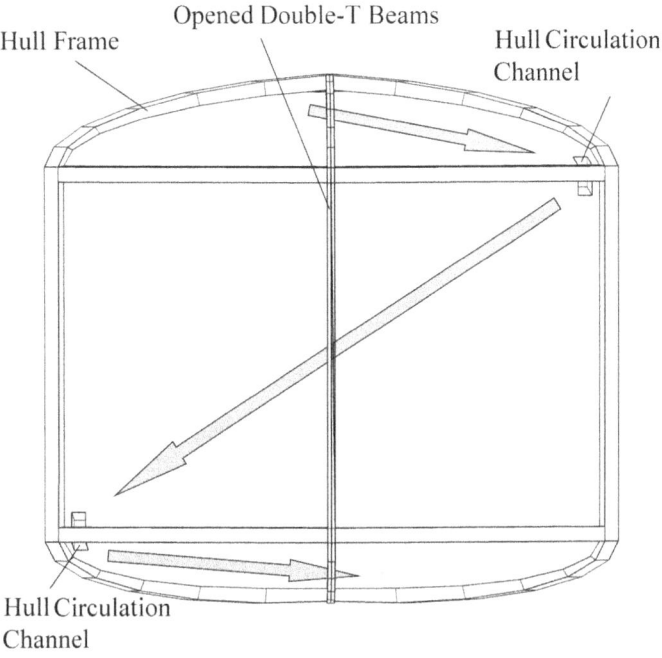

Fig. 4: Circulation of the thermal compensation flow through a spaceship segment

In order to lead the thermal circulation through all eight spaceship segments as closed loop, the single segments are additionally connected to each other in front of the central tunnel wall, so that the circulation flow will be alternately led from the upper to the lower hull side and from the lower to the upper hull side.

The described circulation connections were mounted outside the inner hull planking in order not to influence the stability of the segment frame profiles by inner connection openings. The connection pipes have a diameter of 80 millimeters and are made of a curved aluminum tube. The movement of the outer hull temperature circulation (HTC) is provided by a pump- and ventilation unit in segment 1, which takes there the circulation medium out of the outlet of the lower connection pipe and pumps it back again into the adjoining hull inlet.

2.3.2 Inner Spaceship Walls and Floors

The further separation of the different spaceship areas is provided to the construction by additional lightweight walls. The outer surface of the central tunnel is completely covered with such a wall, whereby the inner projection components are protected from dirtying or damage. Only at the vertical middle of the tunnel, a small opening is located towards segment 1, which allows the passage of the projection light beams.

The spatial separation of the two ring corridors will be made by an inner-sided paneling of the therefore inserted struts. Under the reduced gravitational conditions of a rotation flight, this surface will become an accessible walking floor. In segment 1 this paneling is equipped on the – viewed from the outside – left side with a passage hatch that has a spacing of 0.97 meters upward and that connects the living room with the navigation room in a rotation flight over a further aluminum ladder.

Two inner walls are additionally inserted without supporting structures at each radial side of segment 1. In the outer area of this segment, an L-shaped door has been foreseen that allows the direct access to the planting segment 8. The opening of this door is inserted vertically along the outer wall, but also horizontally along the lower deck surface, so that an upright walk-through is possible during a rotation flight as well as in a landed state. In the inner area of segment 1, horizontally oriented doors are planned to the adjoining segments 2 and 8, which are cut – starting from the upper corner of the ring corridor separation wall – as a lying opening into the radial segment wall. These two doors are intended for use during a rotation flight. In order to make the navigation room accessible also when in landed state, additional vertical doors are foreseen next to the central tunnel. All doors and hatches of the concept spaceship were inserted so that necessary functional units, such as the planting racks, bioreactors, or gas tanks, are not in the walkways and can so be reached very quickly.

The description of the structural spaceship design will be completed with these final design features. If we take the described design as basis, the following material lists can be used as a rough calculation basis. The preferred material for the spaceship construction should be aluminum with the specified material thicknesses.

Table 1a: Material list outer frame

Frame profile 10 x 10 centimeter:	
8 x Main outer half bow à	12.48 Meters
32 x Circular arc sections à	1.95 Meters
4 x Front-pyramid outer edge strut à	4.15 Meters
2 x MCB side strut, horizontal à	4.50 Meters
2 x MCB inner cross strut à	0.32 Meters
2 x MCB vertical strut à	0.30 Meters
Total:	189.08 Meters

Double-T profile 5 x 10 centimeter with openings:	
8 x Main outer half bow à	12.48 Meters
16 x Ring corridor separation wall strut, vertical à	3.21 Meters
Total:	151.20 Meters

with plastic resin in aluminum tubes that afterwards can be airtightly welded with the spaceship hull. Such electrical leadthroughs are foreseen at

- all three landing gear housings,
- the communication antennas in the front pyramid as well as
- for each engine gondola at the outer wall of the segments 2, 4, 6 and 8.

To adjust the air pressure inside the spacecraft, a double-closable outer valve can be inserted at the lower deck surface of the navigation room. Another external valve would be necessary, if the longitudinal tunnel in segment 5 should be used as an airlock and a decompression possibility is required from inside.

The principle of a pressure-tight and plastic capsuled leadthrough is also used to integrate the projection lenses into the central tunnel and the lower deck surface of segment 1.

Another hull opening is foreseen in segment 5 for a dust downpipe with an attached material lock. This downpipe extends from the deck surface of the MCB into the spaceship inner and is – viewed towards the rotation axis – integrated at the front left side of outer segment area. The pipe has a length of approximately 56 centimeters and is two centimeters in diameter. The outer end of the downpipe is located 94 centimeters away from the upper side wall edge and is located directly at the MCB side wall. Below the MCB deck surface, the pipe continues downwards with a 45-degree inclination in outer direction whereto this is fix connected with all crossed hull walls. In the spaceship inner, the downpipe end thus points obliquely towards the outer side wall. The use of this downpipe as well as a solution for a required dust inlet lock are described in the chapter on energy supply and organic matter processing.

2.3.5 Meteorite Protection Fabric

In the component list for the spaceship hull, no planking sheets are listed for the spaceship front-pyramid. This is intentional, because only an aramid fiber fabric will cover this front area. The transmitting and receiving antennas located there, are mounted outside of the pressurized hull, and so they will need only a protection against direct sunlight and meteorite impacts.

The aramid fiber fabric, which protects the complete spaceship hull from meteorite impacts, starts at the tip of the spaceship front-pyramid and extends around the whole spaceship body. It is supported by spacing bolts that have been mounted along all outer edges and curved surfaces of the spaceship hull. Thereby, the distance to the outer hull planking is always ten centimeters. The protecting aramid fabric is firmly closed around the spaceship body and is put under tension, which is able to decelerate impacting micrometeorites before they hit on the outer hull wall. In its function, the aramid fiber fabric absorbs the energy of an impacting meteorite and distributes this by the all-sided increasing fabric tension around the entire spaceship hull (Carrillo *et al.* 2012). Due to its punctual mount and the so enabled fabric surface flexibility, the aramid fabric will be saved from overloads during impact stresses, wherefrom its lifetime will be extended accordingly. The total surface of this protective layer can be assumed for the spaceship body with 247.8 square meters. If the four engine gondolas should be also equipped with this protection, additional 77.3 square meters of the aramid fiber fabric must be taken into consideration.

Since the aramid (aromatic polyamide) is not stable under ultraviolet radiation, the used fabric must be protected on its outside with a light and heat reflecting coating that represents the outermost layer of the hull wall construction (see Fig. 5). An additional advantage of this reflective coating is the avoidance of heating the aramid fabric under sunlight, which is temperature resistant only up to 400 °C. In the hull construction described here, a durable coating with reflective aluminum has been selected for protecting the outer aramid fabric.

2.3.6 Thermal Hull Insulation

The spacecraft hull is provided with a continuous insulation along its inner side to reduce temperature losses as much as possible. For this purpose, a 24-centimeter air-filled spacing has been inserted next to the circumferential outer side wall. Along the upper and lower deck surfaces, this insulation spacing is continued with 20 centimeters thickness. Only in the area below the MCB, the insulation spacing is reduced to 15 centimeters in order to provide sufficient space for all installations underneath.

Also the landing gear housings are surrounded by this insulating layer for reasons of a complete thermal protection. In the front area of segment 1, the front wall planking and the adjoining insulation space form an indentation of ten centimeters in outward direction over an area of 2.125 square meters in order to enable the integration of a low-lying tank for the hydroponic nutrient solution collection. Thus, the insulation is slightly reduced at only a few locations, where supply lines, landing gear housings or the nutrient solution pump need appropriate indentations in the inner hull surface.

Furthermore, occur over those internal construction components that are directly connected to the outer spaceship structure for stability reasons, also an increased heat dissipation. It was taken care to reduce in the construction of the concept spaceship the number of such connections as much as possible and to minimize thereby all additional thermal loss. Therefore, in the calculation for the thermal balance of the spaceship, the expected heat transfer was increased by 63 percent in order to consider also these heat losses. The air-filled spacing was chosen not only because of its thermal insulating properties, it also avoids a further mass increase, since the inner radiation shielding does not need any mechanical support from this layer. For the spaceship concept, the used air insulation has been considered with an average thermal conductivity of 0.026 W/mK at a corresponding density of 1.293 kilograms per cubic meter.

The expected heat loss of the spaceship hull can be exemplary calculated with the inserted air spacing thicknesses and the temperature difference of 218.2 degrees Celsius, which exists between the average outer hull surface temperature in a near-earth orbit and the interior temperature. Depending on the insulation thicknesses (15, 20 and 24 centimeters), the continuous heat energy loss Q per square meter inner hull surface is:

$$Q_{15} = \frac{1 m^2 \cdot 0.026 \frac{W}{mK} \cdot 218.2\, K}{0.15\, m} = 37.97\, W/m^2$$

$$Q_{20} = \frac{1 m^2 \cdot 0.026 \frac{W}{mK} \cdot 218.2\, K}{0.20\, m} = 28.48\, W/m^2$$

$$Q_{24} = \frac{1 m^2 \cdot 0.026 \frac{W}{mK} \cdot 218.2\, K}{0.24\, m} = 23.73\, W/m^2$$

If these heat losses are now multiplied by the corresponding hull surfaces, and if the expected additional losses of the thermal bridges are also taken into account, the calculated heat loss near Earth will be at

$$\begin{aligned} Q_{Tot} = & ((139.8\, m^2{}_{20} \cdot 28.48\, W/m^2) \\ & + (76.48\, m^2{}_{24} \cdot 23.73\, W/m^2) \\ & + (7.2\, m^2{}_{15} \cdot 37.97\, W/m^2)) \cdot 163\% \end{aligned}$$

$$Q_{Tot} = \sim 9{,}899\, W$$

In order to hold the intended interior temperature, this heat loss now consequently must be compensated by internal or external heat-energetic processes, for which reason it is used in the thermal energy balance as a basis demand parameter (see Table 7).

The calculation of the average temperature difference between the outer hull surface and the interior takes into account that high temperature differences can occur over the outer spaceship hull, which are caused by a one-sided solar radiation. During a near-earth orbit, temperatures between +120 °C and -273 °C can easily be reached on sun faced respectively away from sun faced surfaces, which would result in dangerous tensions in the hull construction, if no appropriate counter measures were implemented.

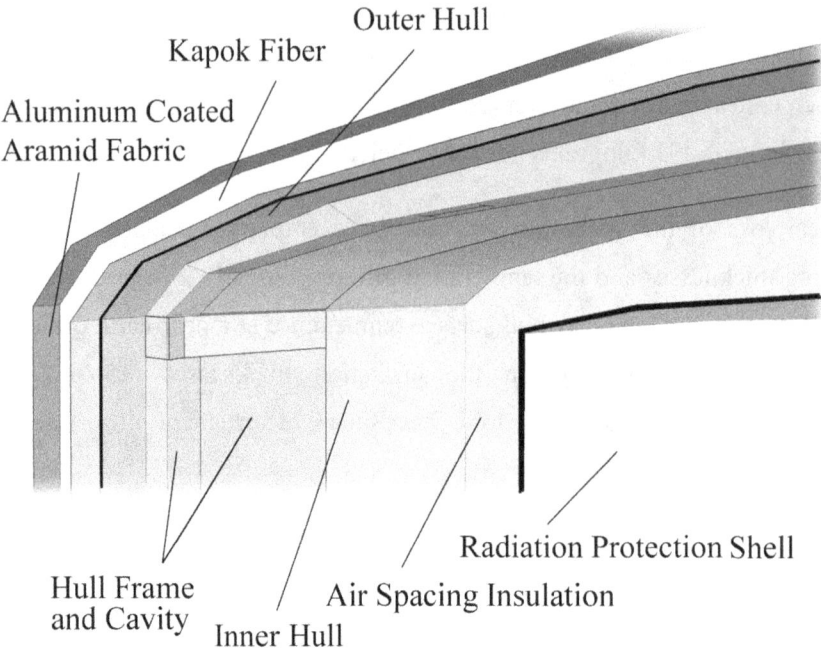

Fig. 5: *Insulation and radiation shielding layers of the spaceship hull*

The reflective coated aramid fabric, which has been placed in front of the spaceship pressure vessel and that has been already described in the context of the meteorite protection, has only at a few small mounting points direct contact with the hull surface. Due to this, it also forms a surface shading that already reflects a large portion of the received heat radiation. Thereby the aramid fabric itself absorbs only a small portion of this heat energy, which is then forwarded in form of an infrared radiation towards the hull wall. In order to minimize also this heat energy flow, the space between the aramid fabric and the outer hull wall will be filled with a heat resistant layer of kapok fibers (Monteiro *et al.* 2012).

At the end, these measures reduce – in combination with the thermal hull circulation flow – the incoming one-sided thermal sun energy of 1,367 watts per square meter to an average of 157.62 watts per square meter, which are uniformly absorbed by the spaceship hull, whereby material stresses from different thermal material expansions can largely be avoided. In addition to the aramid fabric, which for safety reasons should be already taken into account for the spaceship mass, also the mass of the necessary kapok fibers must be considered that were added with a density of twelve kilograms per cubic meter (Voumbo *et al.* 2010).

Due to the outer shading and insulation, the occurring external hull temperatures will be within a smaller bandwidth. Based on the thermal absorption and emission rates of the used shielding materials (Birur *et al.* 2001), a temperature range of maximal -43.5 °C and minimal of -273.2 °C can be determined. If these temperatures are now weighted by the related hull surfaces, which are receiving sun radiation during a flight through the solar system, and those that hereby are in shade, an average outer hull temperature of -200.2 °C will be reached. When the intended interior temperature of 18 °C is now put in relation to this value, the previously used average temperature difference of 218.2 °C can be determined.

The maximum hull surface temperatures under solar radiation suggest for a near-earth orbit that the expected heat loss of a spaceship can be relatively easily compensated in this sun-distance. If in such a position one of the two isolated deck surfaces of the concept spaceship would be continuously exposed to the sunlight, the aramid fabric and the underlying kapok insulation would allow a heat energy transfer of approximately 9,925 watts to the spaceship hull that the thermal hull circulation then could distribute over the entire spaceship hull.

In comparison to a near-earth mission would a mission in far distance to the sun provide much less external energy, so that the thereby increased heat losses must be compensated by a comparable power output of the foreseen high-energy sources, which are described in the chapter on energy supply. If hereto the maximum heat demand is determined, for which the incoming external solar energy is completely taken out from the thermal calculation, the average outer hull temperature would be already lower than the temperature of interstellar molecular clouds (see Herbst 2001). Under these conditions, the so-adapted calculation model of the concept spaceship returns a continuous heat loss of maximal 13,049 watts for the interstellar space.

2.3.7 Radiation Protection Measures

Not to be neglected in the construction of a spaceship are the dangers for space travelers by the extreme radiation in free space. These are far stronger, than they could be absorbed by the previously described insulation and protective layers. In general, a maximum annual radiation dose of 20 Millisievert is allowed for humans, while in the space between Earth and Mars already radiation doses of up to 2.4 Millisievert per day were measured during solar storms (Zeitlin *et al.* 2013). Having such a radiation exposure and being without any additional radiation protection, the maximum annual human radiation dose would be reached after only 8.3 mission days.

In order to shield the incoming high-energy radiation during long-term missions, the spaceship hull of the concept spaceship will be supported by additional protection measures, which have to be dimensioned depending on the possible mission distances and durations. For example, the full-aluminum hull preferred in this document already provides two layers of this metal, which are able to sufficiently shield alpha and beta radiation with their total thickness of 13 millimeters. Thus, the pressure hull wall takes over itself a shielding function, reducing the effort for additional measures.

Much more difficult is the shielding of arriving gamma and particle radiation, which have a significantly higher energy, and for their complete shielding even several centimeters of lead in the spaceship hull would not be sufficient. Detailed studies have been made for the International Space Station (ISS) on this in order to design a protection shielding, which can serve as example for constructing an effective and lightweight radiation protection (Miller *et al.* 2003). The multiple possible shielding variants for the here presented concept spaceship provide as simplest variant a solution similar to this.

Analogue to this design, can the insertion of a stiff, surrounding radiation shielding shell along the inner spaceship insulation spacing – made of alternating layers of polyethylene and plastic-bound graphite – effectively decelerate the incoming radiation. But in order to provide a radiation shielding for the intended long-term missions, several more layers are necessary to increase the protection potential.

In a maximum shielding variant, instead, only a solid polyethylene shell with a thickness of ten millimeters is used along the inner insulation distance.

With its moderate shielding properties, this basic shielding could be directly used for missions to a low-earth orbit or for short space transfers with a limited flight duration. The inner radiation shielding shell, in general, has a surface of approximately 224 square meters, and it is connected to the spaceship hull only by the ring corridor and central tunnel wall structures. The inside radiation shield surface serves, furthermore, as walking surface for the passengers and as a solid mounting structure for all inner components.

Nevertheless, for extended stays in space, the radiation dose of gamma and particle radiation remains a problem, since the previously described shielding measures are not sufficient to spend months or even years in space. Also an increased distance to the sun would not lower the radiation level to be handled, since measurements of the Voyager 1 spacecraft have shown that radiation energies of some hundred mega electron volts (MeV) for protons and up to 1,000 MeV for heavy ions persist also outside the heliosphere (Stone *et al.* 2005).

For this reason, the optionally extended radiation shielding uses the attenuating-effect of water, that enables an effective reduction of radiation energy by the integration of a corresponding shielding thickness. Thereto, the flexible and optional filling of the outer hull wall cavity can equip the concept spaceship with a circa ten centimeter thick water-shell on all sides. Thus, if the priority for a long-term mission is rather on the radiation protection and less on a minimum spacecraft mass, and if water resources could be supplemented also after the spaceship launch (for example with a depot in orbit or at a targeted object), the so inserted water-attenuation would decelerate an incoming radiation energy of one MeV by 50 percent (McAlister 2012). When having the combination of this effective input attenuation and the inner radiation protection shielding, the radiation exposure can be drastically reduced for the space travelers.

This optional water filling of the spaceship hull cavity requires some further technical and mission-related preconditions.

First of all, since the hull wall volume of the concept spaceship encloses about 22.9 cubic meters, a mass increase of 23 tons needs to be considered for its water filling. Comparing the therefrom resulting spaceship mass with that of a concept spaceship using solely the constructive radiation shielding, this would mean a doubling of the total mass. However, the published results of McAlister show that a water shielding provides, in general, the best ratio between mass and efficiency at radiation energies above 0.9 MeV, so that water is the best choice, even for a more capable radiation shield (McAlister 2012).

In a realistic mission scenario, the required amount of water for the hull filling could be launched separately by a standard launching system, like the European Ariane 5 ES, either into a low-earth orbit or in direction of the mission target, whereby the masses to be launched would be divided meaningfully. Furthermore, the water filling can be drained shortly before reaching a destination in order to execute the required braking maneuvers with the lowest possible fuel demand. In the ideal case, the spaceship hull water can even itself serve as a supporting mass for the deceleration by releasing it before the arrival with pressure into the direction of flight. The option of a water-filled shielding can be taken into consideration also for a followed stationary phase, if the filling can be done by using local water resources.

A second topic is about the setup of the required thermal hull circulation. The inserted water can be ideally used to this purpose, for which reason only additional pumps have to be inserted as a supplement to the regular air fans.

Finally, there should be also two risk aspects mentioned related to this additional protection measure. On the one hand, there is a danger from a freezing of the water-filling, which can be caused by a missing solar radiation or an uneven heat distribution in the hull. Since the protecting outer hull shading suppresses almost completely the external heat uptake, the temperature of the hull water would be far below 0 °C having no counter-measures. If thereby the water should freeze to ice, its solidified and increased volume would inevitably damage the hull construction. A possible solution to avoid such a temperature decrease is a parallel use of the shielding water to cool a radioisotope generator, which is described in the chapter on energy supply. Together with the supply of electrical power, this unit warms with its thermal output all the hull water and ensures thereby besides to the transfer of heat to the spaceship inner also its flowability.

A second and certainly positive factor of a water usage in the outer hull cavity is the protection of the spaceship passengers from spontaneous pressure losses, which could be caused by meteorite impacts or mechanical failures. If in such a case a leak occurs in the filled outer wall of the pressure body, first the hull water would be drained through this opening to the outside. Although the necessary water temperature of higher than 0 °C lets not expect an automatic ice sealing of the leak, the higher viscosity of water will increase the time for reaching safe areas or rescue capsules as this leaks much slower than air.

Overall, the variant of the water-filled spaceship hull cavity provides the advantage of better radiation protection, a cooling option for radioactive processes, a temporary seal against leaks, and a general large stock of water, which can also serve as a supporting mass. This is in contrast to a high spaceship mass to be accelerated, what may require two separate launches. In addition, the spaceship hull might be exposed to a potential risk of a freezing that could inevitably damage it.

The concept spaceship allows with its design the flexible usage of several radiation protection options that can be reconfigured also during a mission. For the further concept description, an air-based thermal circulation flow and a radiation protection without a water filling has been selected to allow the spaceship with its reduced engine power also a self-propulsed launch and landing.

2.3.8 Qualitative Hull Checks

In addition to the described mass balance of the spaceship body, also its pressure resistance is of essential importance. In order to ensure both, regular test routines have to be implemented during the production of an outer hull, which will ensure a continuous measurement of the spaceship balance and its ability to hold an inner pressure.

The pressure tests – starting with the completion of the outer hull planking – monitor the high quality of all hull planking connections as well as the correct integration of the hatches and leadthroughs while the further construction stages. Each pressure test is carried out by the introduction of a specific over-pressure, which must be hold during a defined measuring period.

Measurements for testing the balanced mass distribution can be made easiest at the central rotational axis of the spaceship body. Since the spaceship shall rotate around this axis without greater imbalances, all surrounding masses must be radially even distributed. With the circular arrangement of all main components, a basic mass distribution balance will be already provided by the construction. However, for a fine-tuning also all further elements have to be considered in the mass distribution, for which reason the spaceship is lifted at one of the exit points of its rotation axis – either from below by a central support or from above by a chain. If all masses are distributed evenly, the lifted hull will be in an exact horizontal position. For checking this, a pendulum can be used inside the central tunnel that is able to indicate balance deviations very precisely. If an imbalance is detected by this measurement, this can be easily balanced by slightly location changes of the inner units.

The proposed spaceship hull provides with its specified scaling the habitat for one person that now has to be supplied by a life support system. Therefore, in the next chapter the construction of a sustainable life support system and its integration into the concept spaceship are described.

3 Systems for Life Support

As basic components for life support, breathing air, drinking water and food are essential on board of a spaceship. The design of life support for the concept spaceship should be dimensioned so that sufficient reserves of these products are available to supply one or more person. And because there is no way to pick up new supplies during a long space flight, all integrated components for life support have to be recycled in a closed biosphere circulation system.

Such a system must be continuously monitored in order to keep all material cycles in balance. Within a defined parameter range, this monitoring can be taken over by nature, which allows plants and organisms to react automatically on changing environmental conditions. These evolutionary control processes already start directly after the closure of an ecological system, and they influence continuously the composition of all circulating biological components from that moment on. Only just a few hours after the encapsulation, changes of the primary substance turnovers, such as the respiration air regeneration or the drinking water condensation, can be detected. Furthermore, with some delay also secondary parameters are changing in form of growth rates for the inserted plants or microbial life forms, because here a given rate of reproduction comes in addition as limiting factor.

3.1 Plant Species for Space Planting

In previous experiments with different biological life support systems that provided a similar habitat size as the concept spaceship described here, valuable experiences could be already gained in order to allow human inhabitants a survival of up to 90 days under complete materially isolation (Nelson *et al.* 2010). A central component of most of these sustainable life support systems are certainly the continuously cultivated crops, which do not only convert the provided carbon dioxide CO_2 into breathable oxygen O_2 with their photosynthesis, but they also create an eatable and fermentable biomass that is usable for a general nutrient supply. Of course, it is not possible to integrate a large-scale plantation within the limited inner spaceship volume. Moreover, such large plantations would have – if they were be realized with a conventional soil based or a horizontal hydroponic planting – an increased weight due to the required planting medium or large amounts of nutrient liquid.

In order to reach a more efficient integration of the required plant mass, a hydroponic planting variation has been developed, which has advantages in terms of energy demand and the requirements of space and weight. In combination with a selection of suitable plants, a high compressed plant cultivation can so be established within the six planting segments of the concept spaceship, whose high amount of active biomass is crucial for the preservation of the biological balance.

For the plant selection, the following criteria have first relevance:

- High growth rate at low light intensities
- Low growing height and reduced root volume as well as a fruiting on the stalks
- Compatibility of the root system with a ventilated nutrient solution
- Adjusted mineral and nutrient content to supply human persons
- Complete degradability of all plant components and the therefrom resulting feces
- All plant components must be free of poison

Based on these criteria different plant species were observed during several planting experiments in order to test their compatibility to the environmental conditions of the concept spaceship. After a final selection the intended plants species for the basis planting are

- Wheat (Triticum aestivum),
- Spinach (Spinacia oleracea),
- Peas (Pisum sativum) and
- Salicornia (Salicornia europaea).

With their contents of important minerals, carbohydrates, fats and proteins, this combination already provides a good basis for human nutrition. But also other properties such as a good temperature tolerance or the acceptance of a flooded root area supported this decision. For this reason, these plant species times will be in the following viewed from different angles and for several times so that their benefit for the related specific needs becomes apparent.

3.2 Hydroponic Planting

The plantings are located in the outer area of the assigned segments 2, 3, 4, 6, 7 and 8. Within each of these segments, a horizontally pivoted planting rack is mounted that – depending on its orientation – allows the plant cultivation either under an external gravity as well as in the artificial gravity of a rotation flight.

Each of these planting racks has a width of 210.8 centimeters, a height of 179 centimeters and a depth of 78 centimeters. The outer frame of a rack is made of light-weight, L-shaped profiles. In order to enable a pivoting of this frame, it is suspended on its top side with a horizontal axis that extents centrally through the upper rack profile. At both sides, this axis is led out to corresponding bearings that are mounted on the inserted oblique rack holder struts of the IS. The shape of the planting rack frames as well as their mounting positions have been chosen so that they can rotate without any collisions from a vertical into a horizontal orientation.

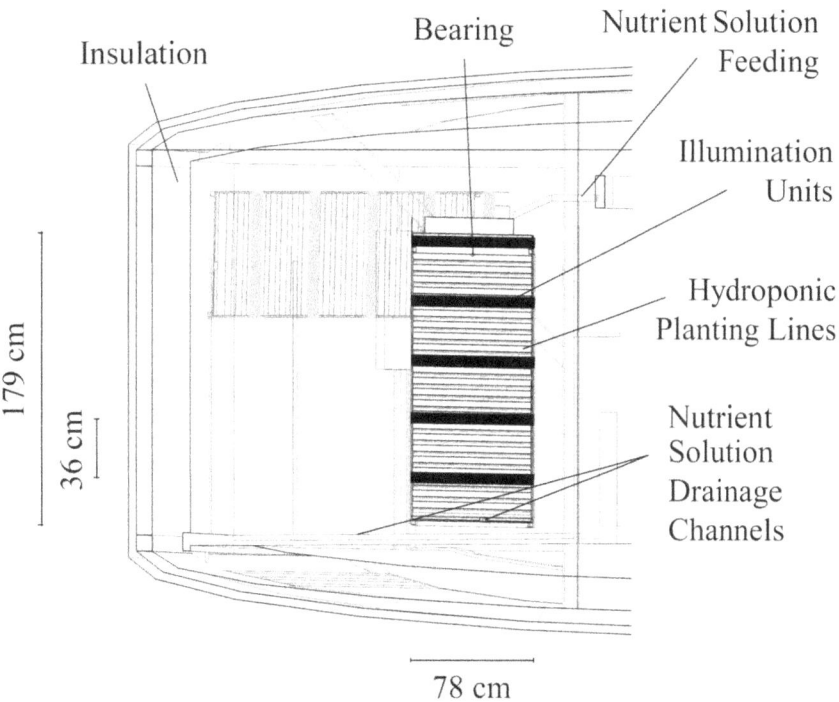

Fig. 6: Pivoting mechanism of the planting racks

The change of the planting rack orientation is triggered by a change of the gravity direction, during which the racks automatically follow with their hanging weight either the external gravitation or the centrifugal force of the spaceship rotation.

For the additional securing of the rack frames also releasable connections to the OHS are foreseen, which prevent unwanted movement during regular flight phases and hold them in a precisely aligned position.

Each of the planting racks includes six upright and longitudinally aligned aluminum or plastic boxes, the so-called planting boxes. Their surfaces facing to each other are provided with special horizontal planting channels that traverse them with a vertical distance of seven centimeters. The upwards opened planting channels serve as a predefined planting raster, give the growing roots a hold, and support the plant growth with a continuously flowing nutrient solution. This will be lead through the planting channels only in little quantities and enables thus a plant cultivation that is done without using heavy soil or large amounts of water, and which represents a combination of aeroponic and hydroponic cultivation techniques.

A planting channel has a total depth of 4.5 centimeters, of which four centimeters protrude on the outside of the planting box. This outer channel section is every ten centimeters divided by root separators into single root areas, which themselves have two millimeter small separators to ease the insertion of new seedlings. The channel has at its front a height of 1.6 centimeters and deepens right up to the planting box surface to four centimeters filling level. The thereto necessary oblique channel bottom one hand helps the plant seedlings underneath to incline into the growing-space. On the other hand, also the volume of the planting channel increases with this design, which is why this has a cross-section area of 11.2 square centimeters at the outside of the planting box.

In order to provide younger and larger plants an increased hold, each planting channel has a continuous support grid at its upper front edge that extends with a ten centimeter wide support surface obliquely upwards. The support grid has large openings so that plant components behind can be supplied with light as well. The top of the planting channel is protected with a removable cover that prevents the growth of algae inside the nutrient solution and only provides a small gap at the channel front-edge for the outgrowth of the plant stalks.

The transport of the nutrient solution through the planting box occurs in a dripping flow from top to bottom. If the solution flows along the inner planting box wall into a planting channel, it will be initially retained there and is lead via sieve-like holes in the box wall to the outwards located plant roots. Since the inner wall of the planting channel is reciprocally lower at one side, the solution can drain there at a certain filling-level downwards to a guiding surface underneath. This oblique surface leads the solution flow back to the inner planting box wall, where it can drip down into the following planting channel.

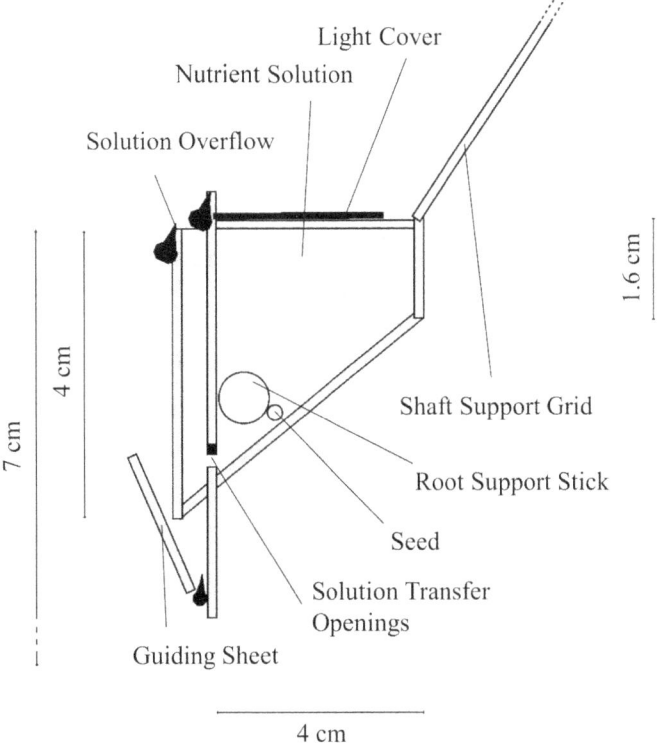

Fig. 7: Cross section of a hydroponic planting channel

3.3 Nutrient Solution Management

The basic components of the nutrient solution are essentially water, added urine and the liquefied and fermented residues from the two so-called bioreactors. In order to produce therefrom a usable nutrient solution, the given mixing ratio is crucial, since the hydroponically grown plants will respond immediately to the

- pH-value,
- oxygen content,
- nitrate content and the
- salinity

of the formed solution. If these characteristic parameters exceed or fall below specific limits, disturbances in plant growth will occur that can lead right to the complete death of all plants.

The pH-value of the nutrient solution is primarily influenced by the amount of the added urine and the released bioreactor substrate. Both additions must be continuously processed within the nutrient solution to hold a constant pH level. The therefrom resulting products are relevant for the plant fertilization and can be taken up by them directly out of the nutrient solution.

Since urine is not helpful to the fermentation process of the bioreactors, it is separated and directly released into the nutrient solution. Therein, a three step process converts the contained urea CH_4N_2O first into ammonium NH_4^+ and then further into nitrate NO_3^-. The chemical break down of the urea is performed automatically by the enzyme urease, which is directly produced by the bacteria within the solution. In combination with the solution water this first step creates carbon dioxide CO_2 and ammonia NH_3. While the formed carbon dioxide will be bound within the nutrient solution to assumed 79 percent in form of bicarbonate HCO_3^-, its remaining portion gradually escapes the solution. Furthermore, the gaseous ammonia dissolves in the solution to ammonium NH_4^+, which increases their pH-value due to the necessary uptake of hydrogen atoms (Vlek et al. 1980).

Also, the daily released substrate of the bioreactors contains – in addition to essential nutrients as, for example, phosphorus – a significant amount of ammonium nitrogen, which moreover will raise the solution content with about 1,025 milligrams per liter (Wang et al. 2012). This additional intake results in an ammonia production as well and causes so a further increase of the pH-value.

If now the pH-value of the nutrient solution rises too high, their uptake capacity for these substances decreases, whereby increasing volumes of carbon dioxide and toxic ammonia gas could escape out of the solution. In order to prevent this in general, the target rate of the in the following described ammonium decomposition has to be enabled at a very high level. By exploiting an additional buffer volume of the nutrient solution, can via this process the increasing pH-value again be neutralized and the release of ammonia gas effectively be prevented.

One portion of the ammonium contained in the nutrient solution can be already absorbed direct by the plants to a nitrogen fertilization. At the same time, a remaining ammonium part is converted to nitrate by two further processing stages. This micro-biotic nitrification process first oxidizes the ammonium by releasing water to nitrite NO_2^-, which then serves as basis for a further oxidation to nitrate (Vymazal 2007). This process is performed by nitrifying bacteria with the involvement of the in parallel dissolved hydrogen carbonate, whereby the ammonium will be chemically bonded, and the pH-value of the solution again will be led out of the alkalic range. Having a continuous ammonium assimilation by the plants and the simultaneous conversion of ammonium to nitrate, a balanced average pH-value will be reached that should be ideally at seven, since this level has proven to be particularly well tolerated by all cereal plants of the planting experiment.

The settlement of the processing bacteria occurs mainly inside the flooded and ventilated root zone of the planting channels, as it happens similarly in so-called constructed wetlands. Such wetlands are used worldwide in the context of wastewater treatment in various designs. Since the planting channels are completely filled with nutrient solution, the root zone of the plants will represent a situation similar to a constructed wetland with free-floating plants (FFP), which is in general characterized by an average rate of nitrification (Vymazal 2007).

With the implementation of an efficient nitrification process, also the analysis of the possible nitrate removal – the so-called denitrification – is required. First and foremost, this certainly must be done by the plants in order to support their growth as much as possible. However, the average denitrification rate of the FFP wetlands also shows that under the given conditions also a bacterial denitrification occurs. This is possible due to low-oxygen areas underneath the waterline, where special bacteria can perform an anaerobic decomposition of the nitrate. Thus, this type of denitrification stands in concurrence with the nitrification, which requires an oxygen-rich environment for the oxidation of the ammonium.

But if the planting channels have to be irrigated in general with a specific air entry, what anaerobic denitrification would then occur under these conditions? In order to determine this, measurements were made to a possible nitrification- and denitrification rate under the controlled conditions of a test planting.

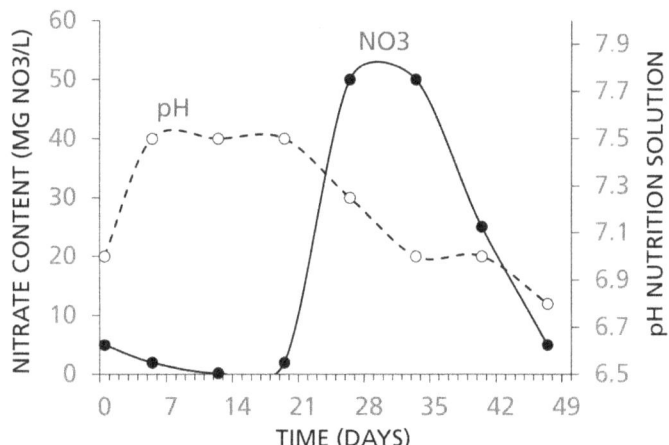

Fig. 8: Rate of ammonium to nitrate conversion and following denitrification

Thereto, an artificially lit and ventilated basin contained a submerged planting channel. All water surfaces of the basin were covered with an opaque plastic plate in order to prevent any growth of algae. The studied nutrient solution should be formed from 40 milliliters of urine, which was diluted by 4.5 liters of water. With the addition of five grams of fresh soil, the needed nitrification bacteria cultures were inserted. Further, an electric air pump aerated the solution at eight times a day for a period of 15 minutes with atmospheric oxygen. Inside the ten centimeters long planting channel three wheat plants were growing that on the one hand provided a flooded root zone for the settlement of the bacteria and on the other hand ensured with their height of up to 30 centimeters a steady uptake of ammonium and nitrate.

Fig. 8 shows the nitrate- and pH-progress of the experiment during a measurement period of 49 days. Initially, after a start-up period of 19 days the expected increase of nitrate content could be observed, which reached its maximum level of 50 milligrams per liter within seven days. In contrast to this, the following denitrification phase required 21 further days, which can be only explained by the low number of plants as well as by a reduced anaerobic denitrification. This leads to the conclusion that even a moderate aeration of the nutrient solution prevents the bacterial denitrification in the root areas, which is why the nitrate assimilation can be done there solely by the introduced plants.

In order to test the concept spaceship for a possible anaerobic denitrification demand initially the daily nitrogen requirement of a single wheat plant was determined from an external study. The nutrient solution used therein had a nitrogen content of 34 milligrams per liter, which consisted equal parts of dissolved ammonium and nitrate (Rroço and Mengel 2000). By the simultaneous supply of 104 wheat plants having a grow age of 18 days, the nitrogen content of 20 liters of this nutrient solution was consumed fully within six days. If these data are now extrapolated to a single plant, thus a daily nitrogen assimilation of 1.090 milligrams can be determined per plant. However, this value does not yet take into account the reduced light conditions of the intended plant illumination, which restricts the nitrogen absorption to the correspondingly lowered growth rate. In general, the growth reduction for the concept spaceship is assumed with 60 percent of the optimal growth rate, which is why also the reference value for nitrogen uptake has been adjusted with this factor to 0.654 milligrams per day and plant.

The nitrogen balance of the concept spaceship contains on the demand side the plants of the 1,028 meter long planting channel. With a relative planting distance of 0.62 centimeters per single plant and taking into account a growth-stage depending demand reduction of 50 percent, their nitrogen uptake corresponds to an equivalent of 82,558 active plants at maximum vegetation. However, two additional factors are influencing this demand: on the one hand, these are the pea plants, whose active portion of 5,865 plants does not reduce the nitrogen content of the nutrient solution. Peas are able to bind the gaseous nitrogen of the air via their aerated roots, which is why they do not require any ammonium or nitrate fertilization. Because of this, the demand determination does not consider them in the number of assimilating plants. On the other hand, the nitrogen demand has to be reduced by further 13 percent due to the determined nitrogen return by Rroço and Mengel. If taking in to account also these two preconditions, a nitrogen entry of 43.64 grams per day is necessary into the nutrient solution to cover the whole nitrogen demand of the plants.

In comparison thereto, the actual daily nitrogen entry consist of four liters of urine and 12.9 liters of released bioreactor substrate that have a nitrogen content of 28.04 grams respectively 13.23 grams. The relative high amount of urine results from an increased human water uptake, which is necessary to make the consumed nitrogen as soon as possible available again for the spaceship plants. Because of this, the drinking water supply has been dimensioned also with the sufficient capacity to cover these daily drinking water needs.

Thus, the nitrogen demand of the plants is with having a slight deficit of 1.47 grams per day nearly balanced from the implementation of a fast nitrogen turnover and the nitrogen-fixing peas, which is why a normal operation does not require an additional bacterial denitrification. But for the case that a greater nitrogen concentration should occur in the nutrient solution – which could be caused, for example, by disturbances in the bioreactor-fermentation, a lower plant growth, or an excessive nitrogen fixation by the pea plants – such a possibility must be present supplementary. Therefore, the concept spaceship also allows a bacterial denitrification inside four nutrient solution tanks, which provide a calmed and anaerobic liquid reservoir that can be used independently from the ongoing nitrification process of the planting channels.

The foreseen implementation of a bacterial denitrification requires, regardless if it takes place with a low rate in the planting channels or within the flexible useable nutrient solution tanks, the presence of easy utilizable carbon compounds in the nutrient solution. These are taken up by the denitrifying bacteria and are oxidized with the available nitrate or nitrite oxygen. Thus, can with this process also organic bioreactor substrate residues be taken out of the nutrient solution, whereby at the same time bound nitrogen will be released into the ambient air. For an efficient denitrification, these organic substances must be continuously supplied to the bacteria, which is already ensured by the foreseen permanent circulation of the nutrient solution.

In addition, all processes of nitrification and denitrification are temperature depending. The thermal balance of the spaceship always has to provide an internal temperature of at least 15 °C in order to achieve the required degradation rate.

Besides the nitrification rate, the pH-value of the nutrient solution will be also influenced from the volume of the circulating nutrient solution, which must provide an appropriate buffer capacity for the ammonium uptake. The total volume of nutrient solution for the concept spaceship overall is 1,428 liters. Out of these, mathematically about 438 liters are held within the planting channels, if the remaining filling volume between the roots is assumed with averagely 45 percent of the channel volume, and the volume displacement of a stick for root fixing is additionally taken into account. Solely this reduced solution volume flows through the six planting segments in a steady circulation, while the far greater portion of the liquid remains in the bioreactors and nutrient solution tanks. This circulating amount of liquid shall now be tested for its potential buffer volume.

Thereto, the already in the planting experiment tested nitrogen content ratio is inserted into a calculation by which the minimum amount of water needed can be determined based on the expected daily nitrogen entry. The used 40 milliliters of urine for the planting experiment contained about 0.4 grams of urea with a nitrogen content of 0.19 grams. After the dilution with 4.5 liters of water, the nutrient solution had consequently 0.0414 grams of nitrogen. This nitrogen was completely processed within 28 days, which corresponds to a processing rate of 0.00148 grams per liter and day.

A significant difference between the planting experiment and the planting channel used in the future is, however, the number of plants involved. In the experiment came 0.67 plants on one liter of nutrient solution, while a planting channel will contain 175 nitrogen-assimilating plants per liter. Considering this, not only the nitrification would run faster due to the increased root volume, but also the available ammonium used for this would be reduced, because the plants will assimilate more ammonium directly out of the nutrient solution. Because this is a crucial factor, the higher number of participating plants has been considered for the calculation as well.

Applied to the total amount of nitrogen the concept spaceship, this means:

$$Nutrient\ Solution_{V\ Min} = \frac{41.27 g\ day^{-1}}{\left(0,00148\ g\ l^{-1} day^{-1} \cdot \frac{175}{0,67}\right)} l$$

$$Nutrient\ Solution_{V\ Min} = 107\ l$$

Comparing this minimum amount of nutrient solution with the real circulating liquid volume, thus a four-time greater assimilation buffer is already available for the in-brought ammonium without any further constructive measures.

The hydroponic planting described here generally requires a good enrichment of the nutrient solution with oxygen, even if the ammonium nitrification is – in comparison to the fuel cells, the bioreactors, or the human oxygen intake – the smallest consumer of this resource. Furthermore, the oxygen inserted for this is relevant for the air supply of the plant roots, which in addition perform a root respiration via the nutrient solution (Bloom *et al.* 1992). Another positive side effect of a good nutrient solution aeration is the prevention of mold growth and other anaerobic rotting processes within the planting channels as well as in the inlet and outlet pipes.

3.4 Nutrient Solution Circulation

The main enrichment of the nutrient solution with oxygen takes place while their mechanical transport and inside the planting boxes, in which the liquid drips downwards from one planting channel to the next. At the lower end of each planting rack, the solution is collected and laterally lead through a common discharge pipe to the outside. The planting segments have a channel system that receives the discharged fluid and transports it towards a collecting channel at the lower outer wall. Since the collecting channel is located there at the lowest point for both possible gravity directions, it can receive liquids under the influence of an external gravity as well as under an artificial rotational gravity. It is designed as a half-open and embedded channel at the bottom corner of the inner radiation shield and it extends – with the exception of the rear segment 5 – through all spaceship segments. Its cross-sectional profile allows the opening and cleaning of the line any time, even if a liquid is conveyed therein.

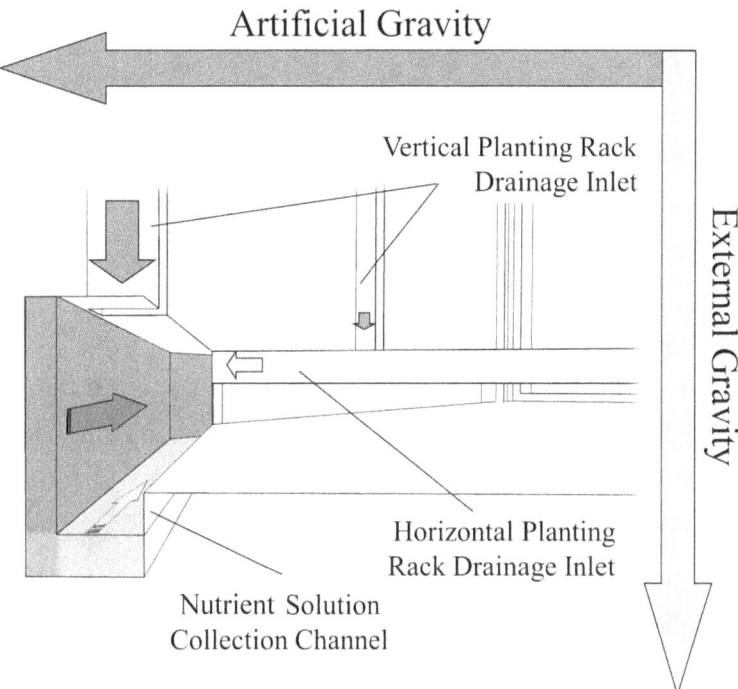

Fig. 9: Cut view of a nutrient solution transport channel for variable gravity

Further inflows in this collection channel occur, if necessary, from the nutrient solution tanks or the two bioreactors, which are located in the inner area of the segments 3 and 7. Their daily outlet in form of fermented biomass and separated urine will be taken up by L-shaped collection containers, which allow the liquid discharge under both gravity directions. The collection containers are equipped with a sieve for filtering out coarse substrate residues and they lead their liquid content directly into the circulating nutrient solution via a connected discharge pipe.

After its collection, the liquid follows the outer wall corner into the living room of segment 1. Here, a central main collection container is located, from which the nutrient solution will be conveyed into the higher lying distribution channel of the inner ring corridor. This will distribute the liquid – after passing a further collection container – by 90 percent to the planting racks and by ten percent to the nutrient solution- and gas tanks. For this purpose, the cross section shown in Fig. 9 can be seen as an example for all other implemented nutrient solution channels, which allow a uniform liquid distribution under both possible gravity directions by using narrow inlet gaps and locking edges.

The ring-formed distribution channel of the inner segment area continues at both sides of segment 1 and leads through all adjoining planting segments. It also passes through a closed pipe the fifth segment that does not contain a planting rack. In direction of the outer ring corridor, single feeding pipes are leading from the distribution channel through the ring corridor separation wall to the planting racks suspended behind. Below the outlet of each of these feeding pipes, the corresponding planting rack has a centered inlet funnel, which guides the supplied nutrient solution via a transverse channel to each planting box.

With the afterwards once again starting passage of the solution through the planting channels, the nutrient solution cycle for the plant supply closes here.

The line system for the nutrient solution includes four nutrient solution tanks (DBT 1-4) that are located on the outer wall of the segments 2, 4, 6 and 8. They contain a variable amount of fluid, which can be adjusted by using a simplified inlet and outlet flow control. Thereby, the nutrient solution flow movement through the tanks can be influenced from a standstill up to a fast tank emptying. The supply pipes out of the inner distribution channel are laid in an arc along the lower shell surface in order not to hinder the pivot movement of the planting racks. Since the tank inlet openings are always located underneath the distribution channel level, the inflow occurs only driven by gravity. Also the liquid outflow is based on this force that drains the nutrient solution, if necessary, into the near located collection channel.

The maximum filling volume of a single tank is 44 liters. A complete filling of all four tanks can so hold back up to 176 liters of nutrient solution for a bacterial denitrification. Since the use of an anaerobic denitrification can be seen merely as an optional measure, and the calculated assimilation of nitrate by the plants would be sufficient during a normal operation, the actual necessary filling level has been assumed with only 50 percent in average. Therefore, the so remaining free storage volume allows a variable adjustment of the nutrient solution level for each of the four tanks, which makes this tank system also usable for the rotation balancing by the navigation and stabilization system. All hereto required rotation balance measurements as well as the general control of the spaceship orientation are described more detailed within the chapter on the space navigation systems.

The nutrient solution flows are ensured by a screw conveyor unit, which lifts the solution diagonally through the outer segment 1 into the inner segment area. By lifting the liquid into an higher lying inner distribution channel, which is located just below the upper deck surface of the spaceship, an automatic gravitational downward flow is made possible under the influence of a given or artificially created gravity. A special feature of the implemented archimedean screw conveyor is its hydromagnetic bearing that provides an abrasion free rotation by using a cone-shaped float and two repulsive magnet assemblies. The main collection container for the nutrient solution, which serves thereto also as a float housing, as well as the pump outlet container have been designed to operate under both occurring gravity directions. The drive of the screw pump is provided by two abrasion free magnetic gearwheels in conjunction with an electric motor that is hydromagnetically beared as well. Details and analyzes on these concept solutions have been aggregated in the chapter on spaceship propulsion.

Since the single planting channels in general will keep their contained nutrient solution, the solution movement in principle can be operated at a slow speed, which is advantageous if using an archimedean screw pump. In order to supply all depending hydraulic functions, a pump capacity of at least 480 liters per hour is needed.

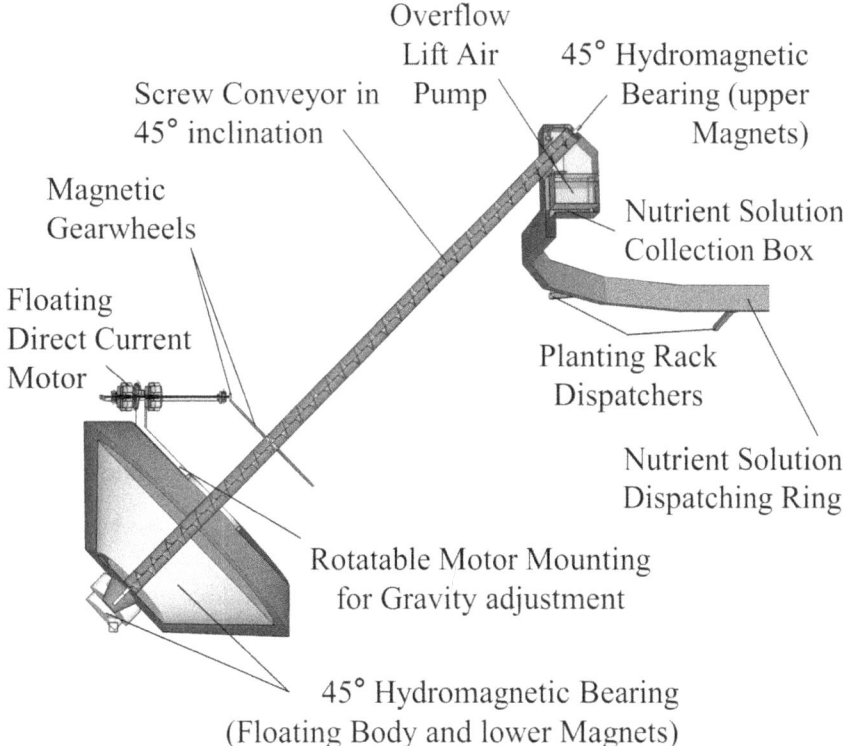

Fig. 10: Archimedean screw pump for pumping the nutrient solution

If the solution pumping has to be disabled temporarily due to mission requirements, this can be compensated quite well by the hydroponic channels, because the liquid still will surround the plant roots. Nevertheless, flight periods with a complete weightlessness must be kept as short as possible, as these not only cause the stoppage of all hydraulic flows, but also increase the risk of losing the nutrient solution out of the planting channels. Taking into consideration that the solution is initially held in such a case by the surface tension of the water and the capillary effect between the roots of the plants, short-term weightless phases are, however, possible. For this, it is only important to pay attention to the avoidance of acceleration forces, which could cause an increased outtake of the solution.

3.5 Illumination of the Plantings

In addition to the nutrient supply, also the lighting of the planting rows is critical to the growth of the plants. Due to the windowless capsule design of the spaceship, no sunlight can be used, which is why an artificial lighting was selected for this. The foreseen light assemblies are located between the vertical planting boxes, whose horizontal distance is 30 centimeters. In order not to hinder the growing plants by a multitude of single lights, only five horizontal light sticks are inserted in each grow space, which are mounted with a vertical distance of 36 centimeters to each other. This vertical lights arrangement ensures a continuous light supply of the plants, even if they grow over one of the light sticks. Moreover, the light of the light sticks overlaps partially in this configuration, so that with an unchanged amount of energy a higher light intensity can be achieved.

A particular challenge for the plant lighting is the fact that the used illumination units cannot be replaced during a space flight. The current state of the art allows the use of LED lights, which in addition to their high life duration of up to 22 years (100,000 hours, twelve hours a day) also have a good energy efficiency (Barta *et al.* 2007). The here preferred stick-shaped design is already available as a series product and allows a good calculation of the required energy demand. This calculation resulted in a permanent illumination energy demand that can be viewed in the electrical energy balance of the concept spaceship (Table 8) in the chapter on energy supply and the organic material processing.

Based on this kind of lamp, the for the planting row lighting best suitable combination of lamp count, lighting distance, and illumination time was experimentally determined. Since the maximum life duration of LED lamps can only be achieved, if they are operated at low currents, also the usage of a LED lighting having a lower wattage was tested. In order to increase generally their provided luminous fluxes, the proposed LED light sticks were combined with a continuous convex lens that reduces the downward light beam angle from 120 degrees to 90 degrees whereby the light will be concentrated to the space between the planting boxes.

During the performed planting experiments, the growth measurement of so illuminated wheat plants allowed the determination of a minimum required illumination intensity and -duration. The lamp configuration integrated in the concept spaceship bases on these experiment results as well as on the therein used LED light sources.

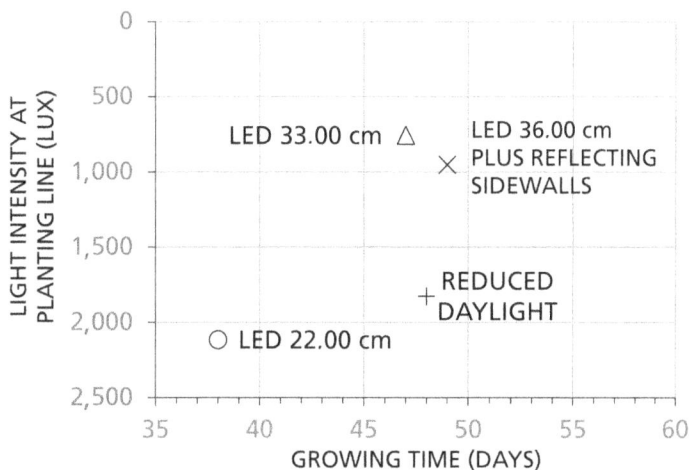

Fig. 11: Diagram to the growth rate, light stick distance and illumination intensity

For a better utilization of the available light intensity, it became necessary to cover all outer surfaces of the planting boxes with a reflective coating. In addition to this, also the grow spaces between the planting boxes will require with a reflecting cover at their front- and back side, which reflects escaping stray light back into the plant area. These covers make it further possible to shorten the length of the light sticks at both sides by 13 centimeters, so that these with their reduced length of 52 centimeters will cause a significantly lower energy demand.

The produced heat output of the illumination is – due to the available LED efficiency of 15.4 percent (reached 105 lumens per watt on maximum technically reachable 683 lumens per watt) – at about 7.15 watts per grow space. In combination with the given gravity, this creates a vertical thermal air movement through the grow spaces, which ensures a constant air exchange for the plants. Furthermore, the emerging heat increases there also the growing temperature whereby the efficiency of photosynthesis is improved (Azcón-Bieto 1983a).

The illumination duration is certainly the most important aspect to ensure a continuous production of oxygen and the removal of carbon dioxide out of the breathing air. In a periodically illumination cycle, the plants need also light-free regeneration times, during which they consume by themselves oxygen and release carbon dioxide via a dark respiration. The oxygen uptake in darkness is at about 0.8 $\mu mol \cdot m^{-2} \cdot s^{-1}$ (Azcón-Bieto *et al.* 1983b), which corresponds to 22.5 percent of the assumed oxygen production by a same sized and artificially illuminated leaf surface. This dark respiration rate has to be taken into account when drawing up an oxygen and carbon dioxide balance.

Another relevant point for the illumination duration is the cycle-depending energy demand, which rises or falls with the number of simultaneously switched-on light sources. If choosing a lighting rhythm, it must be taken care that a uniform energy consumption is reached with it. Therefore, two light rhythms were studied, which are possible with the given six planting segments by assuming a 24 hour day:

Illumination of two segment groups for eight hours

This variant uses three lighting zones, of which only one zone would be illuminated. Each zone would thus have eight hours light and 16 hours darkness per day.

Advantage:
- A lower energy requirement since only 33 percent of the lights are active in parallel

Disadvantages:
- A lower oxygen production due to longer dark periods, which is additionally reduced by a lower starting production at the beginning of an illumination phase

- A shortened vegetative cycle for the plants, which should ideally correspond to the length of a summer day

Lighting of three segment groups for twelve hours

This solution uses two illumination zones, of which one zone would be illuminated. Each zone would thus have twelve hours light and twelve hours darkness per day.

Advantages:
- A balanced oxygen production because of longer illumination times, which reduce in addition the lowered starting production at the beginning of the lighting phases

- A day length comparable with the Earth environment, allowing the plants a longer vegetative cycle

Disadvantage:
- An increased energy demand since 50 percent of the lights are permanently active

Considering the advantages and disadvantages, the twelve-hour lighting rhythm has been selected as the preferred solution, because the thereby provided day length is absolutely necessary for the fruiting of the plants, and the energy demand for lighting can be covered with the intended energy sources.

3.6 Photosynthetic Respiration Air Regeneration

With the selection of wheat plants and all further intended plant species, so-called C3 plants are inserted for the oxygen production, which indeed have a lower maximum photosynthetic rate than the more efficient C4 plants (for example maize), but will have a comparatively better respiration rate under the foreseen ambient temperature and the available light intensities between 1,618 Lux and 9,978 Lux (Rajendrudu et al. 1986). Table 2 shows the measured light intensities of the used experimental arrangement. These are increasing with the reached plant height, which is why the intensities were measured at seven different positions. Cumulative overlays of the vertically arranged light sources are herein already included.

Table 2: Light intensities of the planting lights, depending on the growth height

Growth height (cm)	Cumul. light-intensity (lx)	Light intensity ($\mu E\ m^{-2}\ s^{-1}$)	Portion of illumination zone (%)	Ø O_2 respiration ($\mu mol\ m^{-2}\ s^{-1}$)
0	1,618	20.28	21	1.4
13	3,307	41.44	18	3.3
24	8,262	103.55	3	7.4
26	9,978	125.06	16	8.0
36	1,618	20.28	21	1.4
49	5,784	72.50	21	5.5
Weighted average:				3.6

The biomass involved in the plant respiration can be determined for the concept spaceship by an extrapolation of the physiologically reached plant growth under the given light intensity, the cultivation density, and the spread of the different plant development stages. All needed comparative data of the plants, such as the plant weight and the provided leaf surface per plant, were measured on each three full-grown wheat and pea plants. So reached, for example, the studied wheat plants an average biomass yield of 16 grams per plant at an assumed moisture content of 50 percent. Since the provided light intensity of the concept spaceship reaches in average only six percent of full sunlight, all yield values were corrected to 60 percent of the regular biomass yield. Based on this, the according portions of fruits, leafs and stalks as well as of the roots were determined, which is why these could be included separately in the calculations for plant respiration and the food production.

Finally, the needed root area of these plant species was measured as well in order to calculate a required space-equivalent for the foreseen planting channel cultivation. By using this, it was possible to determine the maximum number of single plants for each plant species within the 1,028 meter long planting channel, which was required in the following for calculating the nitrogen demand, the available leaf surface, as well as the available food components.

Table 3: Plant density and biomass portions

Plant species	Vegetation period (days)	Planting portion (%)	Planting distance (cm)	Daily planting line (m)	Daily yield out of planting line (kg m^{-1} d^{-1})	Biomass total (kg)
Wheat	145	67	0.5	4.8	1.4	865
Spinach	100	18	2.0	1.9	2.4	389
Peas	100	11	1.0	1.2	0.8	86
Salicornia	250	4	3.0	0.1	4.0	67
Averages:		100	0.6	7.9	2.2	1,407

According to these data and assumptions can 1,407 kilograms or 2,896 square meters of permanently active leaf mass be assumed for the respiration processes.

The calculation of the oxygen production is generally based on the measured photorespiration of wheat plants cultivated in space (Tripathy *et al.* 1996) as well as of spinach plants (Osmond 1987). For each brightness zone defined in Table 2, a corresponding respiration rate could be read thereof, which was then used to calculate the average oxygen production rate for each plant species. In a mathematically weighted mean an average oxygen production of 3.56 µmol·m^{-2}·s^{-1} is resulting out of this data.

After the production rates were multiplied by the belonging leaf surfaces, a plausibility check was executed by converting the produced oxygen of each plant species into a daily amount of bonded carbon dioxide, which then was compared to the carbon content of the daily created dry matter. For the preferred wheat plants, thereto an average carbon content of 40.77 percent was assumed. The pea plants were considered instead with 41.93 percent carbon content, which has been set equal to the carbon content of broccoli (Carvajal 2010). Salicornia, in general, has a lower carbon binding, which is why only 25 percent carbon content were used for the plausibility check (Glenn *et al.* 1992). This second calculation enabled thus on the one hand, the check of the assumed gas respiration, but also allowed the adjustment of the expected biomass growth, so that at the end the carbon dioxide turnovers harmonized with the expected biomass yield.

By taking into account all plant respiration occurring with the intended lighting variant, an oxygen- and carbon dioxide balance could be created that served as a calculation basis for all other breathing gas-related processes:

Table 4: Oxygen and carbon dioxide balance

Breathing gas component		
	Oxygen production + O_2 (g h^{-1})	Carbon dioxide uptake - CO_2 (g h^{-1})
Plant respiration (photosynthesis)	+ 559,9	- 769,8
	Oxygen uptake - O_2 (g h^{-1})	Carbon dioxide production + CO_2 (g h^{-1})
Plant respiration (dark respiration)	- 123,6	+ 170,0
Fuel cells	- 396,0	+ 276,2
Bioreactors		+ 275,2
Nitrification / Denitrification	- 6,9	+ 2,5
Human respiration (one person)	- 32,9	+ 45,3
Total:	+ 0,4	- 0,7

Therein, the turnovers of the respiration gases are almost equalized and establish thus a basic equilibrium of the air composition. Additional fluctuations of the carbon dioxide content can be regulated to a certain extent by the adapting respiration of the plants (Azcón-Bieto 1983a). So, for example, already the ongoing harvest of plants triggers such imbalances that are the greater, the larger the harvest areas respectively the required new plantings are. After a plant removal, the resulting carbon dioxide overhang will be first buffered within the ambient air, until the ecological system has adapted itself to the changed gas concentrations. Hereto, a foreseen slight oxygen overproduction in the respiration gas balance eases the adjustment of the life support system to such occurring respiration deficits. Regarding the buffered carbon dioxide it has to be taken care that its portion within the breathing air does not increase to more than four percent by volume, which in such a case would cause a gradual poisoning of the human organism that could finally lead to death, if the portion would rise over eight volume percent.

3.7 Respiration Gas Pressure Changes

The hull of the concept spaceship encloses an air volume of 203 cubic meters, which is held inside at a temperature of 18 °C and a pressure of about 0.8 bar. When launching from the Earth, the internal pressure probably will be first equal to the environment air pressure of one bar, for which reason a pressure adjustment have to be done during the ascent. This can easily be realized inside the navigation room via a valve to the outside, which remains opened during the first launch phase and will equalize the air pressure of the outside atmosphere with the spaceship internal pressure. At an altitude of 1,500 meters above sea level, the valve then automatically closes whereby the correct internal pressure is adjusted.

However, with the closure of the spaceship hull, the internal pressure does not remain static. Changes in the interior temperature – caused by external or internal influences – will any time lead to pressure fluctuations that can be calculated by using the ideal gas law. The corresponding gas pressure of a specific temperature will be considered in the related formula by the molar gas constant of 0.0831447. The required amount of enclosed spaceship air corresponds to a molar equivalent of 6,708.641 mol.

Having these data, the air pressure of the spaceship interior changes within the assumed temperature range between 12 °C (285.15 K) and 30 °C (303.15 K) as follows:

$$\Delta p = \frac{6.708{,}641\ mol \cdot 0{,}0831447 \cdot 303{,}15\ K}{203.000\ l}\ bar - \frac{6.708{,}641\ mol \cdot 0{,}0831447 \cdot 285{,}15\ K}{203.000\ l}\ bar$$

$$\Delta p = \sim 0{,}05\ bar.$$

In addition, the internal pressure is affected by the respective volume proportions of oxygen and carbon dioxide, since these gases have a different molar volume. And also the nitrogen cycle can change the pressure of the breathing air by decreasing or increasing its nitrogen gas content via the fixation of the pea plants or the bacterial denitrification of the nutrient solution. These pressure changes can be determined when comparing a very volume-reduced and an above-average volume-increased air composition. For the volume-reduced variant, a carbon dioxide portion of four percent by volume, a thereto adjusted oxygen content – exemplary caused by the combustion of methane gas in a mass ratio of 1:0.69 (145 percent) – as well as a reduced nitrogen content of 60 percent by volume have been assumed. The volume-increased configuration has instead zero percent by volume of carbon dioxide, a correspondingly increased oxygen content, and a nitrogen content of assumed 85 percent by volume. As before, the base amount of air in the spaceship's hull corresponds to 6,708.641 mol of ideal gas. At 18 °C ambient temperature the pressure difference between these two scenarios now will be:

<u>Maximum Mass Shifts</u>

$$\Delta CO_2 = 6.708{,}641\ mol \cdot 4\ \%$$

$$\Delta CO_2 = 268{,}35\ mol$$

$$\Delta O_2 = \Delta CO_{2\ max} \cdot 145\ \%$$

$$\Delta O_2 = 389{,}10\ mol$$

$$\Delta N_2 = 6.708{,}641\ mol \cdot (85\ \% - 60\%)$$

$$\Delta N_2 = 1.677{,}16\ mol$$

Pressure Difference

$$\Delta p = \frac{\left(6.708,641\ mol - \frac{1}{2}\Delta CO_2 - \frac{1}{2}\Delta O_2 - \frac{1}{2}\Delta N_2\right) \cdot 0,0831447 \cdot 291,15\ K}{203.000\ l}\ bar - \frac{\left(6.708,641\ mol - \frac{1}{2}\Delta CO_2 - \frac{1}{2}\Delta O_2 - \frac{1}{2}\Delta N_2\right) \cdot 0,0831447 \cdot 291,15\ K}{203.000\ l}\ bar$$

$$\Delta p = \frac{7.607,60\ mol \cdot 0,0831447 \cdot 291,15\ K}{203.000\ l}\ bar - \frac{5.509,69\ mol \cdot 0,0831447 \cdot 291,15\ K}{203.000\ l}\ bar$$

$$\Delta p = \sim 0,25\ bar$$

Considering this, the outer hull of the spaceship has to be constructively designed for a pressure resistance of 0.8±0.15 bar, which is why the pressure tests must be carried out with a pressure of at least 0.95 bar plus a safety reserve.

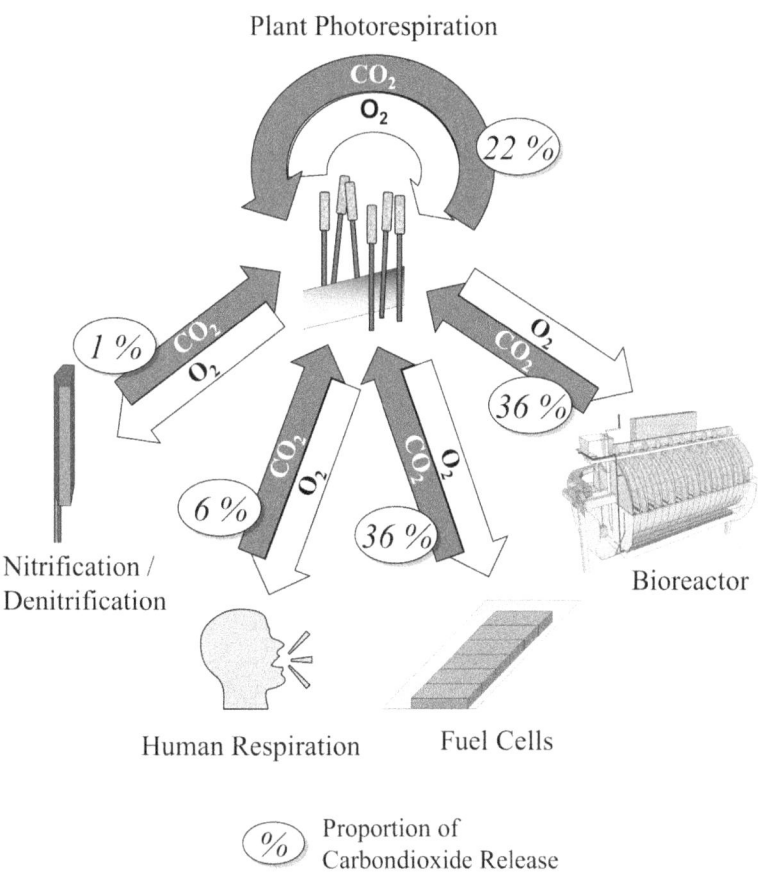

Fig. 12: Influence of the ambient air by the oxygen, carbon dioxide and nitrogen cycle

3.8 Human Nutrition and Planting Management

The plants introduced in the life support system of the spaceship do not only have the task of an oxygen supply and the removal of carbon dioxide from the breathing air, they will also serve for human nutrition needs. It is assumed that a single person has to consume a daily nutrient amount of 2,000 kilocalories out of the running crop production. Since the physical activities of a space traveler in the concept spaceship usually do not go beyond the daily management of planting boxes, various cleaning jobs and light intellectual activity, and because also physical training activities are not necessary due to the presence of gravity, this value – which is comparable to the earthly needs – can be assumed as sufficient.

Table 5 shows an exemplary calculation of the daily calorie requirement of a 35 year old person with 1.61 meters height and a weight on Earth of 62.5 kilograms, who performs all the required activities during a day. The hereto applied calculation method uses the so-called Basal Metabolic Rate (BMR), which corresponds to a caloric requirement per hour – differentiated by age, weight, height as well as by men and women. The different activities are represented by a weighting factor that increases the BMR value according to the physical stress (Henry 2005). The shown example assumes a BMR of 49 kilocalories per hour, which is slightly lower compared to the Earth. This can be explained with the body weight of the previous described person on board of the concept spaceship, which is reduced from the intended artificial gravity to 50 kilograms, which is why the selection of a BMR corresponding to this weight became necessary.

Table 5: *Daily calorie requirement of a person within the concept spaceship (BMR = 49 kcal/h)*

Activity	Demand factor	Duration (h)	Calorie requirement (kcal)
Asleep in bed	1 x BMR	12	588
Management of planting boxes, biomass preparation, maintenance and cleaning	2.8 x BMR	6	823
Data processing, observation	1.4 x BMR	6	412
Total:	1.43 x BMR	24	1,823

All foreseen plant species are generally optimized for high fruit yields, and they usually have relatively small leaf surfaces due to this. Due to this, the integration of a very high number of plants is necessary for the respiration air preparation in order to provide the required photosynthesis performance. Now, thinking about the high mass of daily-available plants, the portion for the human nutrition seems to appear very low. But in fact, only this small portion of the produced biomass is allowed to be consumed directly, as a high nutrient input is also required for the production of biogas inside the bioreactors. Out of the daily harvested 12,896 kilocalories, therefore, about 8,000 to 10,000 kilocalories will be inserted as plant pulp directly into the fermentation units.

Nevertheless, the potential oversupply of food plants opens in this context the option to replace parts of them by inedible, but fast-growing species that can be used solely for an optimized respiration air preparation and biogas production. The daily occurring food surplus of the concept spaceship allows - depending on the nutrition need - a plant exchange with such biomass richer species of between 60 and 80 percent without affecting the intended human calorie supply. The optimized plants will be mechanically chopped after harvesting and will be directly fed into the bioreactors for a further utilization. If the chosen plants have a vegetative reproduction, their use may also reduce the daily effort for replanting. Since only a regular cutting of these plants is necessary, the culturing by seeds thus can be omitted. However, the development of such genetically optimized species is not part of this work, which is why in all further descriptions the sole cultivation of the suggested edible crops is assumed.

All scenarios of a spaceship planting will require a constant renewal of the plant population by a withdrawal of biomass. This is the basis for the respiratory gas preparation described above as well as for all other energetic processes. However, it is mandatory not to exceed thereby over the daily intended planting channel length in order to allow the plants a full growing season and to maintain the balance of the connected ecological cycles. The constructive planting channel length of the concept spaceship is – with having vertical planting channel distance of seven centimeters – already 1,028 running meters, which are spread evenly over the 76.3 square meters of planting box surface. Since the growing season of the foreseen plants, which extends over a time period from the sowing to the harvest of germinable fruits, varies widely from 100 days to 250 days, the daily planting channel length to be replanted must be determined for each plant species individually.

The cultivation with a nutrient solution does not require regeneration phases for a plant substrate, and so the entire planting channel can be used without any time interruptions. During the growing season of a plant species, their assigned planting channel length will be thus once completely harvested and replanted. Overall, the daily planting channel length to be managed for all plant species results in 7.9 meters. Due to the different growing seasons of the used plants, a planting channel management plan has been created for the corresponding planting channel portions that considers the required time intervals in the allocation of the planting channel length and enables a constant daily yield and replanting.

Together with the selection of the plants, also the needed type of pollination has been viewed, since the small spaceship volume does not allow a regular pollination by insects or wind. Therefore, the majority of the proposed plant species has a self-pollination that avoids harvest deficits due to low pollination rates and reduces the effort for a manual pollination. Only the spinach plants (spinacia oleracea) introduced in this concept require such a manual pollination at the time of their flowering. Thereto, the pollen of the male plants have to be transferred with a pollen feather-duster onto the flowers of the neighboring female plants over the daily related 1.85 meter planting channel length. The strong compression of the acreage surface supports this process, since a lot of plant components do already overlap.

After a full growing season, the plants have germinable seeds that can be used both for a food intake and for the cultivation of new seedlings. The fruits and plant material intended for the human consumption can not be cooked for reasons of energy saving and the ensurance of a new planting generation, which is why they have to be eaten in raw and germinated form. This previous germination of the fruits is very important, because it enables a selection of the germinable seeds. Non germinable fruits are thus sorted out from the daily planting, reducing the risk of crop failures.

Since the nitrogen contained in the meals have to be returned quickly into the nitrogen cycle, the daily food must be taken up together with four liters of drinking water. It is also necessary to eat as many of the harvested leaves and stalks as possible, which allows on the one hand, a more comprehensive human utilization of their nutrients, but furthermore ensures their complete fermentation in the following fermentation process.

The composition of the nutrients available for humans is defined by the acreage portion of the used crops and the selection of the thereof required eating quantities. An example to the nutritional composition of a space traveler and the therefrom expected nutrient portions are shown in Table 6.

Table 6: Daily nutrient supply of a space traveler

Plant species	Food portion (%)	Nutritional value (Kcal)	Carbon-hydrate (g)	Fat (g)	Proteins (g)
Wheat	70	1,438	300.3	7.3	38.1
Spinach	15	302	10.7	5.3	53.3
Peas	8	173	10.3	1.8	27.1
Salicornia	7	136	17.2	2.3	10.1
Total:	100	2,050	338.4	16.8	128.6

3.9 Salt Extraction

Another last considered aspect for the nutrition as well as of the nutrient solution management is the collection of salt. Sodium chloride NaCl is very important for the human mineral balance, which is why at least four grams have to be taken up per day and person. The salt is then again discharged into the hydroponic nutrient solution by urination and will accumulate therein. But its concentration in the nutrient solution, however, is not allowed to rise over a certain limit in order to avoid root damages and plant growing disturbances. As a guideline – based on studies of plant growth under different salt concentrations – a salt content of 500 to 1,000 milligrams of salt per liter nutrient solution can be assumed as tolerable (Munns and James 2003). In order to counteract on the increasing salinity of the nutrient solution, therefore, an ongoing desalination process has to be implemented.

The hereto foreseen method uses Salicornia plants, which usually grow along the coastal areas of the Earth. Salicornia has the property to absorb salt via its roots and to store it inside its plant cells, which is why the salt content of a single European Salicornia plant with a fresh mass weight of 180 to 240 grams can reach up to 2.03 grams (Fischer 2016). The Salicornia plant can also be eaten uncooked, and so it covers the human salt needs by a daily intake of about 390 grams. Thus, the great advantage of this form of salt production is that the mineral is directly present in edible form, while alternatives, such as, for example, the boiling of nutrient solution, would always contain components of fecal matter in their recovered residues. The Salicornia plants moreover can adjust themselves to the current salt content of the nutrient solution, so that eventual fluctuations will be automatically buffered by a variable salt uptake. And finally, the dried biomass of the Salicornia plants can be used for the production of soda by being burnt to ash and washed out with water. This process results an alkaline electrolytic solution that directly can be used after its filtering for the experimental fuel cells of this concept. Further information about these fuel cells have been summarized in the chapter on energy supply.

Since this salty plant thus serves to two different functions, their cultivated plant portion needs to be adjusted accordingly. If assuming 50 percent of the human need for an ongoing electrolyte preparation, the daily required harvest quantity of these plants can be calculated on the basis of the given data:

$$Salicornia_{g^{-1}day^{-1}} = \frac{6\ g\ day^{-1}}{2.03\ g\ plant^{-1}} \cdot 200\ g\ plant^{-1}$$

$$Salicornia_{g^{-1}day^{-1}} = 591\ g$$

Assuming further a vegetation period of 250 days and a required planting channel length of three centimeters per plant, a total planting channel length of 36.9 meters has to be foreseen thereto, whereby the daily harvest and replanting of five Salicornia plants will be enabled. This plant number has been already mathematically adjusted by the reduced growth rate of 60 percent of the normal yield due to the reduced light intensity.

The operation of the planting racks is generally done in manual handwork by one or more space travelers. Certainly could this activity be mechanically supported due to the uniform designed planting racks, but provides the daily maintenance of the planting channels – in addition to the saving of weight and energy – also a mentally and physically balancing activity in the otherwise not very varied life on board. The required routines and duties fill thus a good portion of the spaceship day whereby an elementary sense is given to the daily life. In general, the therefrom arising situation can be compared with the nostalgic adventure story of Robinson Crusoe, who also maintained each day the fields of his island and so ensured for himself a certain level of supply.

3.10 Drinking Water Condensation

Another important topic to the life support of the concept spaceship is the supply with drinking water. The temporary storage of the drinking water is done by four identical tanks, which are located in the segments 2, 4, 6 and 8 on the spaceship center facing side of the ring corridor separation wall. Their balanced distribution around the rotation center of the spaceship reduces thus the remaining imbalance during a rotation flight. In total, the tanks have a capacity of 28.6 liters, which in general seems to be very low. However, new drinking water is produced continuously by the spaceship systems, which is why these tanks serve only as a small temporary storage. The tanks – named as clear water tanks (CWT 1-4) – are connected to each other via interconnection pipes along the ring corridor separation wall, so that after a water removal to an outer segment, their filling levels will be automatically balanced over the principle of communicating pipes.

On the one hand, a drinking water removal can be made via a branch from the tank interconnection pipe in segment 1, which leads into the adjoining living room. Here it can be taken out via a tap for the human needs. By using a second branch, the pipe extends here even further in outside direction to provide the water also to the neighboring floating direct current motor that drives the nutrient solution pump and the ventilation system. Another water extraction is made in segment 4, where a supply pipe leads to a second floating direct current motor, which is used in the outer segment area solely to generate a static charge.

The water produced in the concept spaceship serves to a defined number of purposes. These are in particular

- the human drinking water supply,
- the bearing of the floating motors,
- the production of electrolyte for the fuel cells as well as
- the optional supply of further processes of the high energy generation or the space engines.

Its production starts with the oxidation of hydrogen, which is obtained from pyrolytically treated methane gas. The thereto required fuel cells are inserted in the segments 3 and 7 underneath the bioreactor containment, where they are submerged in an electrolyte solution. Since the resulting water is inserted directly into the electrolyte solution, their water content increases continuously. But at the same time, the created oxidation process heat of 240 watts increases the electrolyte temperature to averagely 60 °C, which is why a large amount of water evaporates from the given electrolyte surface of 1.74 square meters. Both bioreactors are equipped with their own electrolyte shell, so that the water vapor now can be used to warm the each contained substrate by being guided upwards along the fermenter wall. Finally, the vapor enters two condensers that are located on both sides of each bioreactor. These condensers are fix connected to their bioreactor-fuel-cell-units, which is why the thermal-driven vapor inflow can take place also after a change of the gravity direction as soon as the unit has been rotated into the corresponding upright position.

For a stable temperature level, the bioreactor-fuel-cell-units are provided with a ten centimeter thick insulation made of polystyrene foam. It also encloses the 1.54 meter long electrolyte shell and forms so the necessary cavity for the vapor guidance along the arched fermenter wall.

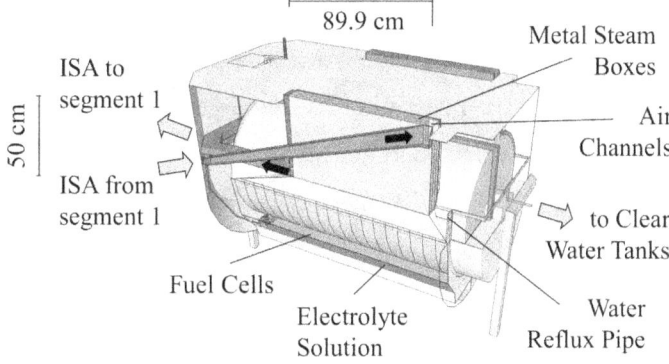

Fig. 13: Cut view of two water condensers with an underlying bioreactor-fuel-cell-unit

Each condenser has itself three flat chambers, in which the supplied vapor can condense on cooled surfaces. At the lower end of each condenser surface, a collecting channel is located via which the obtained water will be discharged. The condensation surface of all four condensers comprises 10.79 square meters. In order to cool the condensation chambers from the outside, these are placed within an air stream that is provided by the ventilation system for each condenser. Since the condensers itself are not insulated at their outside, they provide a small portion of the heat energy directly to their ambient air as well. On the basis of executed condensation experiments, a daily produced drinking water volume of 31.6 liters could be calculated for all condensation surfaces of the concept spaceship. Thereto, the results were extrapolated that have been measured with an average steam temperature inside the condensation chambers of 53 °C, a cooling air temperature of 18 °C and a cooling air supply of 2.7 cubic meters per capacitor and hour.

However, most of this water amount will be lead back into the electrolyte solution, since the limited water production of the fuel cells allows only 4.3 liters per day to be taken out therefrom. This amount of drinking water will be gradually discharged over the day from the condensers and drips laterally out of the bioreactor-fuel-cell-units into the neighboring drinking water tanks.

3.11 Air Circulation Management

The cooling air of the condensers leaves the bioreactor-fuel-cell-units with a temperature of 47 °C. In order to use this for a general air warming, it will be released into segment 1 for an air circulation through the outer ring corridor. The hereto designed aero-thermal circulation system consists of two separate air circuits. A first circuit serves to the temperature compensation inside the outer spaceship hull walls, for which their inner cavities were provided – as described in the chapter on the outer hull construction – with a circular flow path. This thermal cycle can be realized optionally with air or water. In addition, there is also the option to completely deactivate the hull circuit if the spaceship receives no or only weak sunlight and the temperature differences along the hull will become so low, that no thermal compensation is required. The second circuit is implemented for the respiration air, which continuously flows through the spaceship inner and thereby does not only support the heat distribution, but also mixes the transported gases and distributes them among the gas producers and gas consumers.

For this purpose, a central ventilation unit has been inserted in the outer segment 1, which is driven as well by the floating direct current motor of the nutrient solution pump. The air movement is there provided by two direct-driven radial fans, which at the upper motor axis are mounted within the motor suspension. Since this motor and its flexible suspension must be able to follow the gravity direction and to rotate around the axis of the nutrient solution pump, all input and output air channels of the fans are connected by circular air channels and a rotatable cover around the nutrient solution pump.

When using an air-based thermal hull circulation flow, the hull air will be sucked out of the outer wall in segment 1 at the lower circulation connection and be pushed back by the ventilation unit into the lower segment deck surface.

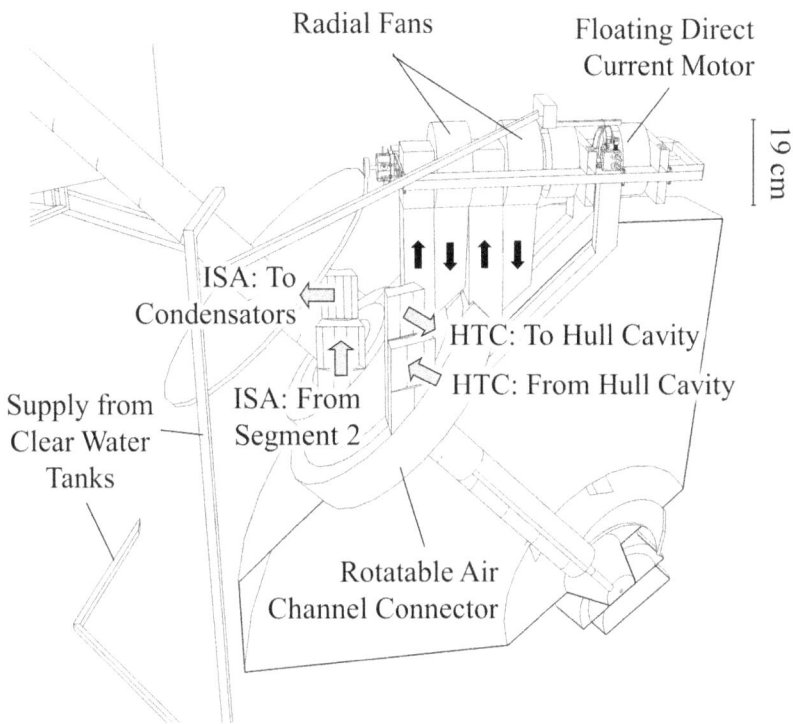

Fig. 14: Cut view of ventilation unit with floating direct current motor

A second, inner air circuit uses the ambient air from the outer segment 2, which the ventilation unit sucks through the radial separation wall of the living room. Starting from the outlet of the ventilation unit, this air stream is first passed via a channel to the navigation room. There, the channel branches and extends to both sides along the ring corridor separation wall to the bioreactor-fuel-cell-units. The coupling of the air channels to these pivotally mounted reactor units is made laterally of their suspension via rotatably covered connection channels that enable their free rotation. After the condensers of the bioreactor-fuel-cell-units have been passed, the warmed air is in parallel to the feeding channels returned to segment 1, where it reaches the living room over an outlet opening through the ring corridor separation wall. The inner air circuit now uses the spatial separation between the segments 1 and 2 for the further airflow control. In order to reach again the inlet of the ventilation unit, the air must now either flow along the outer ring corridor through all planting segments, or it takes an alternative path over the navigation room and segment 5, where it additionally passes the bioreactor-fuel-cell-units.

Both airflows are of the same importance, since the inner segment flow supplies the fuel cells with oxygen for the methane and hydrogen oxidation, while the outer segment flow provides the thereof resulting carbon dioxide again to the planting segments for the photosynthesis.

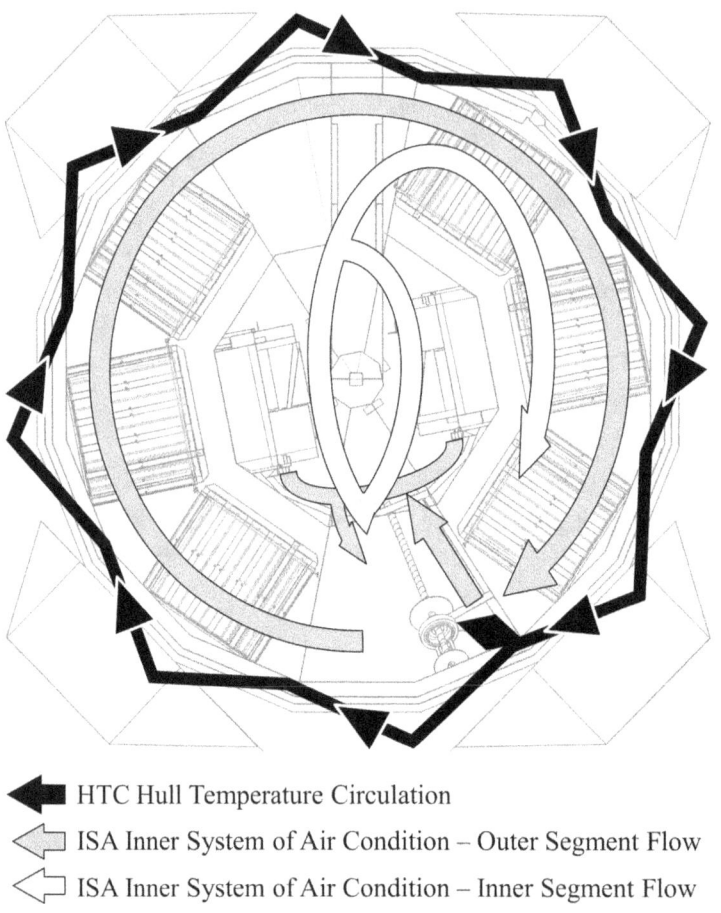

⬅ HTC Hull Temperature Circulation
⇐ ISA Inner System of Air Condition – Outer Segment Flow
⇐ ISA Inner System of Air Condition – Inner Segment Flow

Fig. 15: Air flows in the concept spaceship

4 Systems for Energy Supply and Organic Material Processing

4.1 Internal Energy Supply

All components of the here described concept spaceship need energy in form of electricity, which basically enables the operation of lighting, motors or the navigation. But before this can be generated, first the consideration of another part of the nutrient cycle is necessary that acts as an energetical and biochemical counterpart to the implemented growing of plants. For that reason, the objectives for the energy supply include – next to the production of heat and electricity – also the processing of waste materials out of the plant cultivation and the human nutrition.

If looking closer to the energetic processes on board of a spaceship, so starts the resource based energy production already with the digestion of uptaken food. An adult person provides thereby a total power output from 95 to 300 watts at an average workload. Although the heat portion of this energy can be already added to the thermal balance of a spacecraft, this output is, however, not really usable as energy source for the intended electrical consumers. Shall now a more sufficient energy-generation process be added, so inevitably the further processing of the already consumed food and all other vegetable substances has to be taken into consideration. The uptaken food is excreted after its digestion via urine and faeces that still contain a plenty of nutrient components, which can serve directly as a basis for a biochemical decomposition. Excluded from this decomposition is only the urine that is – as described in the chapter on life support systems – directly introduced to the nutrient solution, where an aerobic nitrification process makes the contained nitrogen faster usable for the plants. But it is a different situation with the collected feces and the large amount of daily harvested biomass, which consist of mostly solid components that must not be introduced into the nutrient solution circulation. And furthermore contains this material a variety of usable nutrients, which is why the energy generation process starts exactly here.

4.1.1 Bioreactor Units to the Degenerative Fermentation

In order to obtain a most homogeneous substrate for the biological degeneration, as much plant material as possible – including also leaves and stalks – has to be taken up by the space traveler. During the human digestion, their components will be already biochemically decomposed to a certain level and can therefore be easier processed in the following. All other plant materials are mechanically fine-chopped and are fed into the fermentation process as well, so that the whole biomass of the plant cultivation can be used for energy generation. The calculation model of the concept spaceship indicates here a daily plant fresh mass preparation of 12.9 kilograms per day.

The substrate to be fermented will be alternately inserted into the two separately located bioreactor-fuel-cell-units of the inner segments 3 and 7. In order to ensure the function of these units under the influence of an external gravity as well as in an artificial generated rotation gravity, they have been suspended in parallel to the ring corridor separation wall with a central rotary axis.

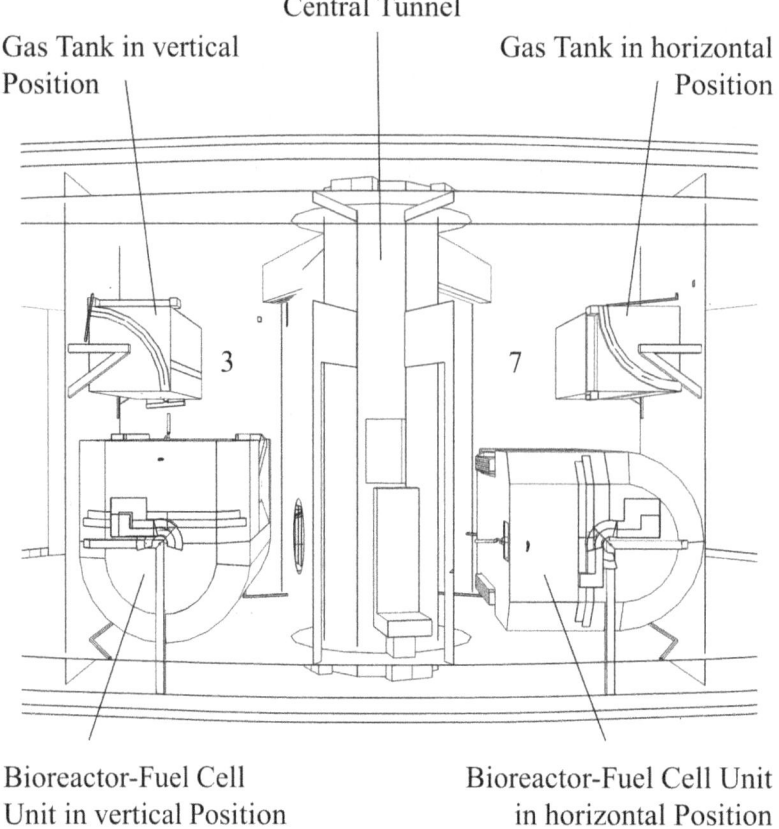

Fig. 16: *Arrangement of the bioreactor-fuel-cell-units*

The bioreactor units have at their front end an upper feed opening, which also serves as a toilet. However, since the radius of the inner ring corridor is relatively small, and the units occupy most of this area, the space for using the toilets is very scarce. Particularly if the units are in their rotation flight orientation and shall be used in a sitting position, this will require the usage of the two doorway openings to the adjoining segment 1, which in this case provide a sufficient headroom.

Inside the feed opening two separate discharge zones are located, which automatically make a separation between faeces and urine. The connected urine pipe leads from here directly to the opposite end of the unit, where it drains the liquid into the collection box for the nutrient solution. All solid faeces as well as the biomass are supplied to the bioreactors over the second discharge zone. The substrate will be thereto mixed with a daily total amount of 12.9 liters of nutrient solution and is then pressed by a screw conveyor through a vertical pipe under the inner liquid level to the bottom of the reactor. Both reactor container same like a lying cylinder that is feeded at one side and is emptied at its opposite end. The reactor cylinders have a diameter of 87.5 centimeters, and their lower half is divided into 19 single fermentation shells, of which each is 8.1 centimeters wide. If these shells are completely filled, each bioreactor can hold a volume of 451.4 liters. Taking both units together, 902.8 liters of fermentation substrate thus can be processed simultaneously. Above the fermentation shells a horizontal screw conveyor is placed that fills out completely the diameter of the reactor cylinder. The screw is provided with vertical gaps and extends with its lower half into the single fermentation shells. By a half turn, each of the immersed screw section transports the content of a shell into the next followed shell. Thereto, the screw conveyor has been also equipped with longitudinal separation sheets, which support on the one hand the shifting of solid biomass components out of the fermentation shells and on the other hand will be a cover for the shells after each half turn.

To operate the feeding screw as well as the highly translated horizontal screw conveyor, a hand crank is attached next to the upper feed opening, which enables the manual driven forwarding of the reactor contents. If necessary, the big horizontal screw conveyor can be decoupled from this drive if only the feeding screw shall to be operated.

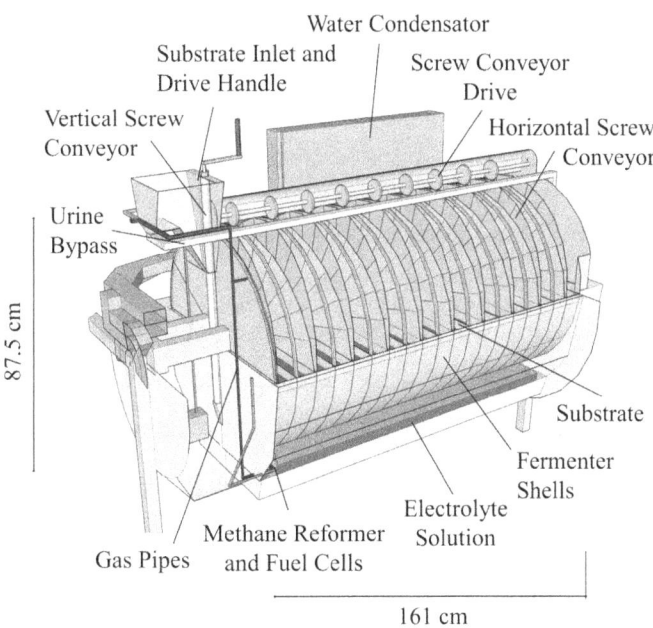

Fig. 17: Cut view of a bioreactor

All fermentation shells can be individually detached downwards for maintenance. If residues block the further transportation of the biomass, hereby the necessary cleaning can be done without emptying the complete reactor content. The described shell coverage by the longitudinal screw conveyor sheets prevents thereby a larger air inlet into the upper half of the reactor cylinder.

The cylinder end lying in the outlet direction has in the following of the last fermentation shell a fall chute with a closing shutter. Any time, new biomass is inserted through the feeding pipe, the reactor releases here automatically a corresponding amount of already processed substrate. Considering the provided daily biomass and the available reactor volume, this substrate previously needed 35 days to go through all fermentation shells.

The anaerobic digestion that starts with the filling of the bioreactors passes till the substrate discharge through four process steps in which different bacterial cultures are involved:

1. Hydrolysis of the nutritional components by which these are decomposed into generally smaller components,

2. Acidogenesis that converts the hydrolyzed components to hydrogen, carbon dioxide, acids, and alcohols,

3. Acetogenesis, which transforms the prementioned substances to acetic acid,

4. Methanogenesis that does the final conversion to methane gas and carbon dioxide.

The fermentation occurs solely in the absence of oxygen, which is why all filling- and discharge operations, by design, are executed without an air insertion. In addition to the absence of oxygen, three further main factors define the efficiency of the bioreactors: the pH-value, the temperature, and the quality of the fermentation substrate.

During the fermentation, there is an increased risk that the pH-value of the substrate drops too much by the formation of acids. When reaching a pH-value below 6, the methane-producing bacteria cultures would be already massively inhibited. However, this can be prevented by a slight supply of urine, because the thereof resulting alkalic ammonia can compensate the acidification of the bioreactor. Ideal is a pH-value between 7 and 8.5, which should be kept during the fermentation process (Linke *et al.* 2006). The spatial subdivision of the bioreactors into individual fermentation shells assigns clearly defined areas to the fermentation steps whereby the mutual influence of the bacterial cultures is reduced. Nevertheless, allow the shells their partial mixing during the substrate transport, which is useful to the continuous bacterial inoculation of the substrate.

In order to determine the expected total biogas yield, the possible yield has been extrapolated for each type of biomass within the calculation model of the concept spaceship (Linke *et al.* 2006). Based on a daily supplied biomass of 12.9 kilograms and an experimental fermentation temperature of 35 °C, the calculated biogas yield accordingly amounts to 193.7 liters per hour. However, with its methane gas content, this amount of biogas is not sufficient for the continuous operation of the intended fuel cells, which is why the gas yield must be further increased by additional measures.

As first, a good mechanical shredding of the not eaten biomass can provide here a better nutrient availability during the fermentation. In the meanwhile, the fermentation shells are warmed by the steam guided along the outside of the reactor container, which will allow to keep an inner temperature of about 55 °C. This enables the usage of thermophile methane bacteria that in this temperature range will provide an improved gas production from which the produced biogas contains an average methane portion of 50.5 percent. Together with an optimized biomass feed of 0.54 kilogram dry mass per liter reactor liquid, not only an increase of the biogas yield by 56 percent can be expected, but also the required processing time can be shortened hereby, which is why the two bioreactors – even with their limited reactor volume – can process the upcoming biomass within the available time (Bouallagui *et al.* 2004).

In summary, the so improved biogas production meets the requirements of the intended fuel cells and provides thereto a biogas amount of 302.1 liters per hour. Since this mainly contains carbon dioxide in addition to the methane gas, also a carbon dioxide emission of 275.2 grams per hour has been taken into account for its production inside the oxygen- and carbon dioxide balance.

4.1.2 Biogas Storage Tanks

The biogas produced from the fermentation requires a slight positive pressure for its distribution to the connected consumers. This can be achieved by the continuous production of gas, whereupon the needed pressure is build up by the closed bioreactor design. Supplementary thereto, also the gas pipes leading to the fuel cell units terminate underneath specially submerged inlet funnels, which define the maintained pressure, since the gas has to be pressed here through an outlet below the electrolyte liquid level.

Besides the direct use of the biogas inside the fuel cells, this is also led on demand into two connected gas tanks that are located in direct neighborhood to the bioreactor-fuel-cell-units. These have a rectangular shape and are – following the gravity direction – also rotatably attached to the ring corridor separation wall. The tanks consist of a square side surface having an edge length of 54 centimeters, and their horizontal width extends over 158.5 centimeters in parallel to the wall. Both tanks together have a biogas storage volume of 0.924 cubic meters at only a minimal compression, so that already the gas supply pressure to the fuel cell units is sufficient to fill them. This feature avoids the use of gas compressors, which is why the biogas storage can be done without any electrical power demand.

For a variable storage volume, the gas tanks have a movable piston sheet in their inner, which is pushed by two spring-toggle mechanisms underneath with a very low force against the upper surface of the tank. If now a gas storage occurs, the inner piston sheet will be pushed down in order to provide a greater storage volume. On the other hand, if the supply pressure drops and the gas in the tanks will be needed again, the piston sheet moves upwards instead and presses so the biogas back to the bioreactors and fuel cells. This principle is similar to an earthly gasholder, only that no gravity is used to generate the storage pressure, which is why the biogas pressure can be held even during a weightless flight phase. For sealing the movable tank piston, a small amount of inflowing nutrient solution is used, which distributes over the entire piston sheet surface and which trickles laterally through the circumferential fugue at the tank walls. In order to prevent a gas leakage via the nutrient solution inlet, the solution coming from the distribution channel is first passed through an U-shaped siphon at the side of the tank. The discharge of the liquid occurs via a small opening in the downside tank surface, which is equipped with a weight-based closure valve that automatically closes the outlet with a spring-lever as soon as the gravity force disappears.

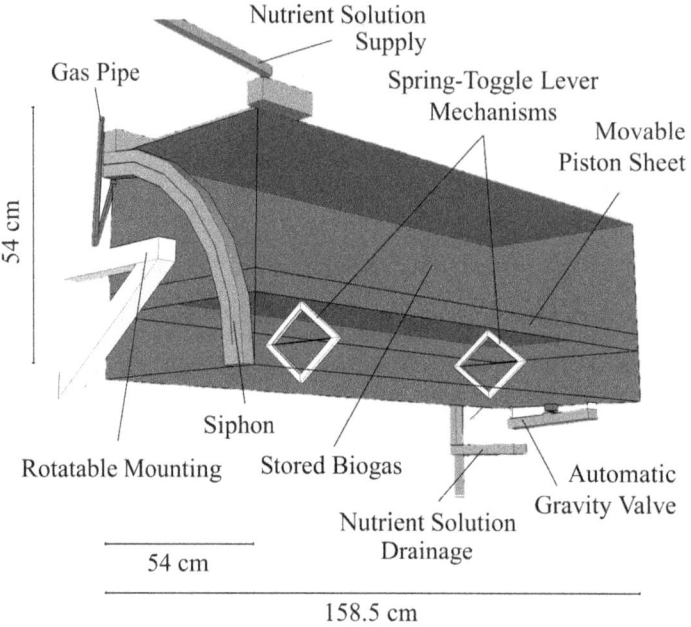

Fig. 18: Cut view of a biogas tank

Ideally, the gas tanks will be used only to compensate an over- or underproduction. With having an optimal adjustment of the fermentation process, the amount of consumed biogas should be equal to the gas amount produced, which is why the gas tanks will keep then a constant fill volume. To increase the gas reserves, therefore an overproduction of biogas has to be initiated by, for example, a temporary shut down of the computer system. This will cause an electrical load reduction at the fuel cells, which in the following also results in a lower biogas consumption for electricity generation. As a consequence of this, the introduced biogas will be accumulated within the submerged fuel cell inlet funnels. By pressing down so the electrolyte level to the gas inlet, the required supply pressure now continuously increases, which then causes the automatic storage of the surplus gas. If, on the other hand, the supply pressure drops by a higher gas consumption, the available gas storage pressure is sufficient to press additional gas under the liquid level into the inlet funnels instead.

Together with the long-term storage of biogas, a portion of the gaseous carbon dioxide will be bond out of the respiration air. A complete filling of the storage tanks will increase the air oxygen content via the upstream biochemical processes by about 611 grams, while the carbon dioxide content is lowered by 1,680 grams therefrom. This general shift in the respiration air composition can be only compensated by the consumption of the stored gas.

4.1.3 Methane Gas Reformer Units

The gas volume required for electricity generation has been determined by a reference to a conventional hydrogen-powered fuel cell product. If this is used as an example to generate the internal electrical energy demand according to Table 8, a hydrogen volume of about four liters per minute would be required (see Schunk 2007). Since the fuel cells of the concept spaceship are also based on the oxidation of hydrogen, the available biogas respectively its methane gas content has to be first reformed into hydrogen gas. Stoichiometrically, there could be two parts of hydrogen H_2 produced out of one part of methane CH_4, so that the biogas demand could be calculated by taking simply the half of the hydrogen demand. But because the biogas produced in the bioreactors has merely a methane gas content of assumed 50.5 percent, which in addition can be reformed by the following described reformation process for only about 80 percent, the needed biogas demand has to be increased by these unusable gas amounts. Taking this also into account, an hourly needed biogas volume of 150 liters can be assumed, which is covered by the projected gas production of the two bioreactors.

The conversion of the available biogas into electrical energy completely takes place in the electrolyte shells underneath the bioreactors. The electrolyte contained therein consists for the fuel cells of the concept spaceship of a solution of sodium carbonate Na_2CO_3, which is obtained by the combustion, leaching, and filtration of the cultivated Salicornia plants. The solution thus becomes electrically conductive and has a mean pH-value of 9.

Due to the released oxidation heat of the fuel cells, the electrolyte bath will be continuously heated, which is why it should reach – in interaction with the fermentation- and condensation processes – a constant temperature of 60 °C. This so balanced temperature level now shows the great advantage of the combination of bioreactor, fuel cells, and condensers, which allows the direct and simultaneous use of the available oxidation heat for a biogas production, the generation of drinking water, and the warming of the spaceship air.

The exemplary referred fuel cells enable an electrical voltage of 24 volts at a power output of 15 amperes. The two-staged oxidation of the thereto used methane gas, which is first converted by the reformers and then oxidized in form of hydrogen within the fuel cells, requires an hourly oxygen supply of 396.0 grams, while it releases in the same time 276.2 grams of carbon dioxide into the ambient air. As a part of the concept calculation, these values were taken into account accordingly within the oxygen- and carbon dioxide balance.

The mentioned carbon dioxide release is mainly generated by the reformers, which bind the carbon atoms of the methane gas CH_4 to the oxygen atoms out of the additionally introduced water steam H_2O. This type of methane decomposition is necessary, as the materials for the fuel cell electrodes and the electrolyte cannot directly oxidize the methane gas under the given temperatures. In order to obtain instead the much easier to oxidize hydrogen H_2, the concept uses simplified reformer units that consist essentially of a vertical nickel tube through which a centered resistance wire runs that has a high nickel content as well. The resistance wire is heated with about six watts of electrical power to a dark red glow. This heat is then transmitted by the produced infrared radiation also to the surrounding nickel tube, which has for this purpose a small inner diameter of 5.1 millimeters and a very thin wall thickness.

The lower end of the nickel tube is connected to a funnel, which is submerged with its wide opening into the fuel cell electrolyte. Inside the reformer funnel, thus a liquid level is formed of warmed electrolyte that lets constantly rise water steam into the funnel, which then is led together with the inserted methane gas into the reformer tube. The methane gas is supplied underneath the electrolyte level via an elongated and likewise downwards opened outlet funnel. This type of gas feeding establishes, on the one hand, a simple flame arrester, while on the other hand, the gas output pressure in the outlet funnel serves to a general pressurization of the biogas, which is necessary for a gas storage inside the foreseen gas tanks. Moreover, can a immersed gas outlet be used as a suction barrier for an upstream overflow lifter pump, which is why such a submerged inlet funnel has been inserted also into the supply line for the fuel cell air.

Over the reformer tube, whose reactive nickel serves as a catalyst to the cracking of the methane gas, a metal cap is placed that takes the upwards exiting hot gas stream and leads it down again along the outside of the nickel tube. This counter-current flow first provides a greater heating of the nickel tube, so that its efficiency will be increased accordingly. Onto the inner surface of the counterflow cap, which provides only a narrow gap around the nickel tube, iron(III)-oxide Fe_2O_3 has been applied that is used for a further reforming of resulting carbon monoxide CO.

The diameter and the operation current of the resistance wire have to be dimensioned so that a sufficient mechanical stability is maintained and a burn-through of the wire is prevented. Therefore, the resistance wire has been based on a wire diameter of 0.1 millimeters and a passage resistivity of 143.9 ohms per meter. Since the biogas flows only at a low maximum speed of 2.52 liters per minute through the reformer tube and is furthermore led by the narrow flow channels closely to the heated nickel metal and the iron oxide of the counterflow cap, each reformer will require only an effective heated resistance wire length of 6.3 centimeters.

By the heating of the biogas-water steam mixture to about 800 °C, the desired reformation process starts. Under the given oxygen exclusion, the introduced methane and water steam molecules will be cracked and will be reformed into carbon dioxide and the required hydrogen, which then can be processed directly in the subsequently connected fuel cells. The mixing ratio of steam and biogas should ideally be at 2:1, for which the reformer funnel needs to collect approximately five liters of wet steam per minute. Since this type of methane reforming produces also toxic carbon monoxide, a so-called water gas shift process has been added inside the iron oxide-coated counterflow cap, which converts in a further process step the carbon monoxide and the remaining water steam to carbon dioxide and additional hydrogen (Alves *et al.* 2013). At the end of the reformation process, the mixture out of steam, hydrogen, and carbon dioxide is led out at the lower end of the reformer cap via a lateral pipe connection.

In order to reach a high efficiency for the gas reforming, in general the provided temperature, the supplied quantity of steam, the rate of gas flow, and the length of the reformer tube and the reformer cap need to be adjusted properly. Since the amount of inserted steam is already defined by the size of the inlet funnel and the electrolyte temperature, and also the expected gas flow through the reformers is set by the gas demand, it was only necessary to determine the required minimum length for the resistance wire, the nickel tube, and the counterflow cap via a supplementary experiment. The aim was hereby to use a shortest possible resistance wire length in order to keep the required energy input for the reforming process low. In particular, because each bioreactor-fuel-cell-unit will be equipped with its own counterflow reformers, and therefore two of these units have to be supplied with electrical power at the same time, this optimization was necessary.

Having the oxidation of the hydrogen completed by the fuel cells, their outgoing air- and vapor stream still contains unreformed methane gas that has been removed again from the cells by a specific cell membrane permeability. In order to oxidize also this gas amounts and to prevent therewith an increase of the methane content in the respiration air, the reformers take over another function in addition. After leaving the fuel cells, the exhaust stream is initially bubbled underneath the electrolyte level to another submerged funnel, which serves as a second flame arrester and furthermore causes the separation of all contained liquid. Then, the gas mixture is led via a pipe to the top of the reformer counterflow caps, where it will be ignited by the upper end of the glowing resistance wire that is led out there by two millimeters. The complete post-oxidation continues then laterally from the counterflow caps inside combustion tank whereby the so obtained heat energy can be directly provided to the reforming process.

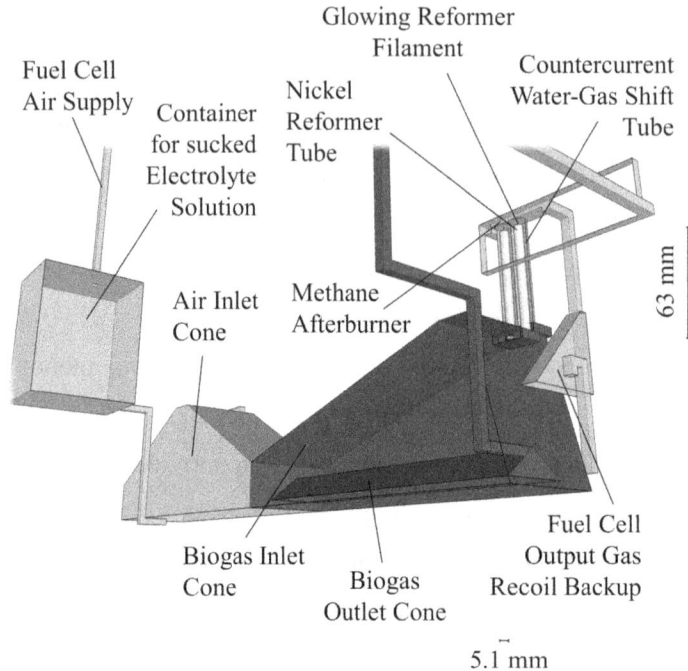

Fig. 19: Counterflow reformer with inlet funnel for hydrogen production

After this last oxidation step, the exhaust gas contains a high portion of carbon dioxide that results from the biogas production as well as from the reforming process. In order to prevent local carbon dioxide over-concentrations, the exhaust gas cannot be released direct at the bioreactor-fuel-cell-units, which is why it is passed through a pipe into the outgoing cooling air channels of the condensers, where it becomes a part of the inner air circuit through all spaceship segments.

In connection with the reformer description, three further concluding aspects of the biogas preparation are to be mentioned. A general danger from the biogas reforming is the conversion of methane components to unwanted by-products. In particular, the carbon monoxide with its toxicity can be risk to the entire spaceship ecosystem. Only with the insertion of the described water-gas shift process, the carbon monoxide gas can be largely bound, which is why this process can be optimized, if necessary, by the length of the used reformer components.

Furthermore, can the inserted carbon portion be bound by a so-called reformer coking, in which pure carbon atoms deposit onto the surfaces of the nickel catalyst and the counterflow caps. Such deposits can be generally reduced by a higher steam saturation of the reformer gas. If nevertheless such a coking occurs, this can be simply burned off by taking the reformer out of the regular operation and flushing it with only air. The glowing resistance wire then ignites the surrounding carbon whereby this will be burned to carbon dioxide and becomes so again a part of the respiration air.

Biogas feeded reforming systems often have problems from a contamination of the supplied gas with hydrogen sulfide H_2S that quickly can disable microscopically treated catalyst materials and is also highly toxic. In order to avoid also these risks, the iron oxide as well as the used nickel metal are preferably used in not treated or activated form, whereby their performance will be although greatly reduced, but their potential use time instead can be extremely extended. In general, both catalyst metals are insensitive to the used liquids and gases and cannot be further corroded by the reactive oxygen. However, ideally the hydrogen sulfide will be bound already before it reaches the reformation process by the electrolyte solution, as it dissolves in water and disintegrates gradually due to the supplied air. Thereby, pure elemental sulfur is formed, which is why the solution has to be completely exchanged at regular intervals.

Table 7: Thermal energy balance

Thermal component		
	Loss	(W)
Heat loss of the spaceship hull		- 9,899
	Production	(W)
Heat energy out of biogas reforming		+ 12
Heat energy out of fuel cells		+ 240
Heat radiation of electric consumers (according to table 8)		+ 287
Heat generation of a human (70 kg, 1.70 m) at an average of 12 hours resting and 12 hours light activity (Havenith *et al.* 2002)		+ 149
Inner thermal balance:		- 9,211
	Additional uptake	(W)
Optional heat with sunlight on spaceship hull		+9,925
or		
Optional supply of heat from radioisotope generator or fusion reactor (10% of 12,000 W)		+ 1,200
Maximum thermal energy overproduction:		+ 714

In the context of the thermal energy balance for the concept spaceship, the preparation and oxidation of the biogas has been taken into account in different ways. At the beginning, the methane gas reforming at the nickel catalyst remains allotherm and provides only a small thermal output from the heat radiation of the resistance wire. In contrast to that, the followed water gas shift reaction offers already a slight exothermic energy output from the oxidation of the carbon monoxide (Alves *et al.* 2013). If, in addition, also a reformer decoking is required, the initiated cleaning combustion can generate a low heat radiation as well. However, the fourth and largest heat release takes place inside the fuel cells from the oxidation of the formed hydrogen by which also the last components of the supplied methane gas are stoichiometrically reacted. Therefore, the complete heat energy content of the processed methane gas can be inserted into the thermal energy balance. Furthermore, must be therein also the heat output from the consumed electrical energy included that is resulting from the operation of the plant illumination, the reformers, and all other electrical equipment.

4.1.4 Overflow Lifter Pump

The transportation of the hydrogen gas to be oxidized into the fuel cells is performed solely by the production pressure from the bioreactors and the enlarging gas volume of the methane reforming. But for the fuel cell air supply, however, an overflow lifter pump is required, which has been inserted underneath the outlet of the nutrient solution pump in segment 1. After the liquid has flowed out of the solution pump, it is collected in an L-shaped container above the inner nutrient distribution channel. By reaching a defined filling level, an overflow lifter then empties the liquid again out of the container into the underlying nutrient distribution channel. Due to the chosen L-shape, the installation of two overflow lifters in a 90 degree angle has been made possible whereby the pump operation can be assured in both possible gravity directions.

Inside the so automatically filled and emptied pump container, another L-shaped container is placed, which is closed on all sides and has only two small, lateral liquid-compensation openings in its lower respectively outer corners. If now the nutrient solution level of the pump tank rises, the liquid also pushes through the compensation openings into the inner container whereby the therein enclosed air will be compressed. At the inner corner of the pressure container, a pressure line has been attached that leads the compressed air to the bioreactor-fuel-cell-units, where it will be bubbled into the submerged air inlet funnels. Because the connected air outlets are thereto located below the electrolyte level, these also serve as suction barriers, through which only the positive air pressure is passed to the fuel cells. This becomes relevant as soon as the outer pump container empties and the falling liquid level causes a negative pressure within the inner pressure container and the thereto connected pressure line. In order to oppose this suction force with an enlarged blocking mass, supplemental electrolyte containers have been integrated into the pressure line, which laterally of the electrolyte shells accumulate the sucked electrolyte solution and prevent thereby a further sucking into the pressure line.

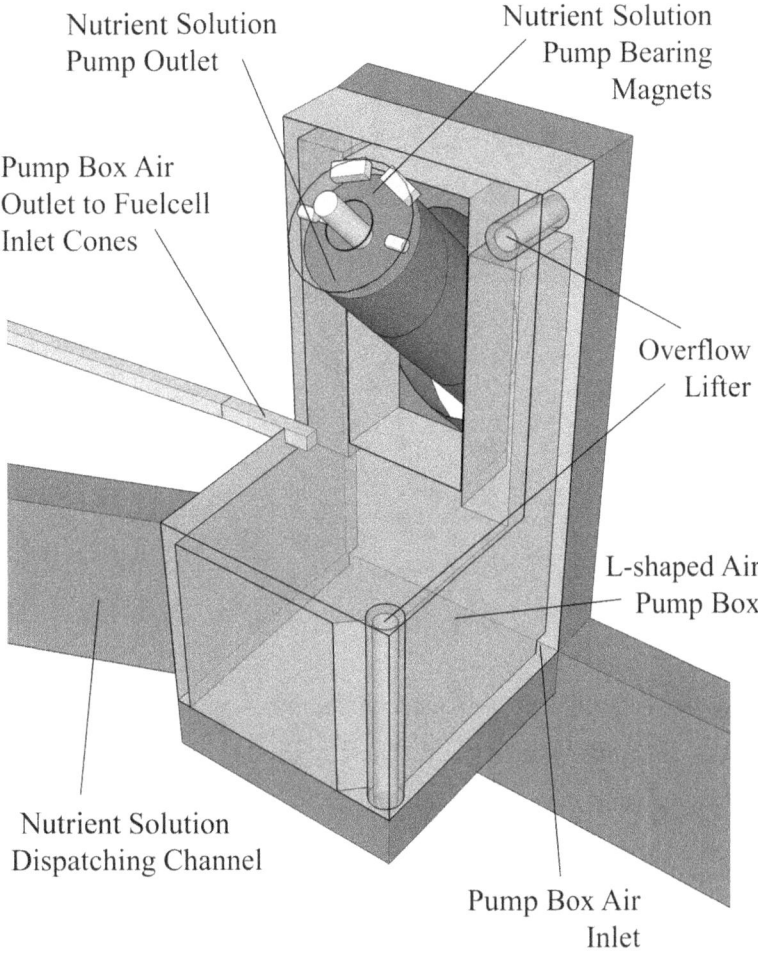

Fig. 20: *Cut view of the overflow lifter pump*

While the overflow lifter pump empties its liquid, the inner pressure container can receive air only first when the liquid level has fallen to the compensation openings, whereupon this empties likewise and will be refilled with air for the following pump cycle. These pumping operations as well as the reformer functions described before require mandatorily the presence of an external or artificial generated gravity that, first of all, enables the needed water and steam movements. Also for this reason, the general provision of a gravity has been foreseen, under which these components can be integrated without problems.

4.1.5 Simplified Hydrogen Fuel Cells

The fuel cell units consist of a combination of numerous individual cells. In order to extract the required voltage and current, their outputs will be accordingly aggregated by serial or parallel connections. In comparison to the referenced fuel cell *(Schunk 2007)*, the concept spaceship can provide a 183-times enlarged cell volume to generate the same electrical output power of 360 watts. This scaling is required due to the simplifications in the cell design and the therewith related greatly reduced power generation efficiency. Each individual cell therefore has a surface measure of about 24.3 centimeters width and

9.3 centimeters height. The actual reaction surface therein is slightly reduced, since a circumferential frame with one centimeter width has to be taken into account.

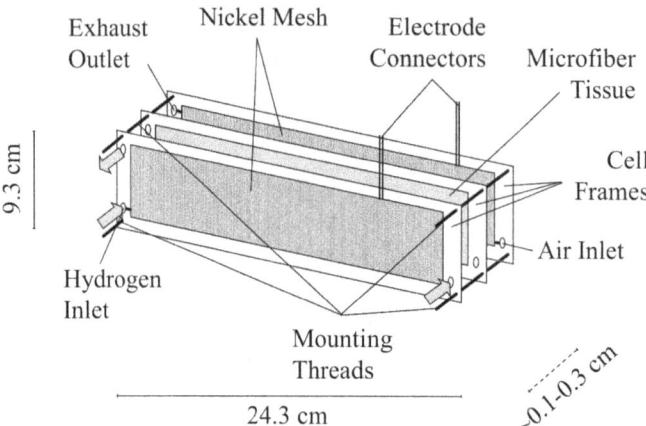

Fig. 21: Single cell of a fuel cell block

Modern fuel cell designs generally strive to a strong compression of their components in order to reach a maximum energy efficiency. Thereby, highly porous membranes are usually used, which were coated with platinum catalysts for an easier decomposition and oxidation of the supplied gases. These membranes are also especially designed for the transfer of ions to ensure a strong flow of electrons through the overlying electrodes (Kim *et al.* 1995). And last but not least, do this kind of fuel cells use strongly alkalic or acidic electrolytes, which support the efficiency of the oxidation by their high conductivity.

In contrast to this, the fuel cells of the concept spaceship – analogue to the simplified production of the electrolytic solution from water and the ash of Salicornia plants – are also made of insensitive and not very specialized materials. Although this reduces the performance of a single cell, it also simplifies the cell design that much that, if necessary, a maintenance and repair is made possible with the on board available tools. The use of simplified membranes between the gas areas provides already a coarse separation of the fuel gas from the oxidizing air, but it allows in parallel to the needed ion flux also an undesired flow of electrons that effects like an internal short of the cell. However, physically this negative property can be improved by a greater distance between the gas areas, which means a higher thickness of the membrane. In the experiments to this simplified fuel cell design, a plastic microfiber cloth was used for this purpose that in a moistened state was able to separate two adjoining gas areas physically. The thickness of the cloth was about one millimeter. Onto both sides of the microfiber cloth woven nickel wires were applied, which can – as an alternative to platinum catalysts – likewise trigger the decomposition of the reaction gases (Mund *et al.* 1977). These electrodes were laterally led out of the submerged fuel cell whereto their cables had a complete electrical insulation in order to prevent further short currents. The nickel wire electrodes of the cells did not undergo any special treatment, only a cleaning and a slight mechanical roughening of their surface was made before they were used.

The so constructed experimental fuel cell was finally enclosed with a rubber frame and two plastic cover plates, which allowed the control of the gas flow to the electrodes. All screws through the cell frame were tightened with only a low torque, so that the microfiber membrane was able to take up liquid from outside, whereby a capillary driven exchange of the inner electrolyte solution has been enabled. For this reason, a separate electrolyte supply via the inlet gas pipes became no longer necessary. The gradually introduced electrolyte is then automatically discharged during the cell operation via the outgoing air- and exhaust gas flow. Because the fuel gas diffuses continuously in ionized form through the membrane, it requires only a single supply line. On the side of the oxidation medium, instead, a continuous air supply has to be implemented, which is why for this gas area separate inlet and outlet lines have been foreseen. The schematic single fuel cell of the concept spaceship shown in Fig. 21 has this gas routing already integrated in its single cell frames, whereby automatically – solely by the stacking of the cell block – branching supply lines are formed that will finally provide the three described main terminals at the end of the cell block.

The internal electrical energy balance, which bases on the available fuel cell energy, should be dimensioned to enable the operation of the life support, the navigation systems, and the on-board computer. This comprises thus also the supply of the plant illumination, the air- and nutrient movement, the methane reforming, the external material collection as well as all basis instruments and the sensors connected thereon.

Table 8: Internal electrical energy balance

Electrical component			
	Power generation	(V)	(A)
Fuel cells		24	+ 15.0
	Power consumption	(V)	(A)
Demand for pyrolysis		24	- 0.5
Plant illumination		24	- 12.5
Conveyance of air / water		24	- 0.3
Operation of MCB generator		24	- 0.3
Computer systems, navigation and sensors		24	- 1.4
Max. energy overproduction:			0.0

4.2 External High Energy Sources

In addition to the internal energy needs, of course, there are other electrical requirements, like the movement of the engine gondolas or the operation of the spaceship engines. Since these functions are not immediately relevant for the life support, are not necessarily based on electrical power, and their energetic requirements change frequently during a space flight, they will be considered separately in the following. A special attention is hereby paid to the all-time availability of the external power supply, which is a mainly relevant criterion for the space engine components. In particular, the changes between rotated and unrotated flight phases can only be ensured if the engines are independent of the internal energy generation, which is based on the presence of a gravity and therefore ends after a short time in weightlessness.

The kind of external power supply is generally depending on the type of propulsion engines, which can be chosen differently for the concept spaceship according to the mission profile and budget. In the case of chemical propulsion engines, the required large amounts of fuel and oxidant have to be stored in an externally connected propulsion unit, since their volume and their weight would exceed the intended capacities of the spaceship design by far. For this purpose, appropriate external propulsion units are already available with a sufficiently large capacity and from various providers of launching systems, so that at this point the function of rocket drives will not be discussed in more detail.

These days, chemical propulsion engines are mostly used for transfers to objects in the near neighborhood of Earth. Their contribution to a mission usually ends with the achievement of a predefined orbit or flight course. In connection with these near-earth missions, also the usage of chemical navigation thrusters would be conceivable for the concept spaceship, which are based on the catalytic decomposition of hydrazine or its reaction with an oxidant. Their relatively simple technique can easily be added – together with the smaller fuel tanks – onto the circumferential side walls of the spacecraft as it is realized also for other spacecraft (Barber and Cowley 2002). The additionally required electrical energy for these engines is limited to the control of the individual hydrazine valves and a mandatorily required tank heating and -ventilation, which can be covered, if necessary, also by the capacity of the internal energy supply.

In the chapter on the spaceship engines, also alternative approaches are described that will operate solely on an electrical or mechanical basis whereby the fuel-based supporting mass can be avoided. Nevertheless, this should not obscure the fact that – even with a minimum spaceship mass – still very high thrust forces must be provided by the used engine systems. Only so can launches, landings, or course corrections be performed with a sustainable acceleration. The external energy supply required for this is primarily from electrical or thermal energy, which must be generated in very large quantities whereto the spaceship mass, however, is not allowed to be increased significantly. In order to narrow down the field of research for such high-loadable energy sources, three approaches have been considered, which were individually analyzed for the concept spaceship. Which of these solutions can be most usefully implemented, depends in general on the state of the available technology, the mission duration and the foreseen operation area.

4.2.1 Solar-based Energy Generation

As a simple and classic power generation form that already has a very good operational history, the solar cell has been established. The advantages of this proven and well researched energy source are certainly the good availability of all necessary components as well as the favorable prices. Moreover, it would not be a technical problem to attach rotatable solar panels to the outside of the spaceship body that could be permanently aligned over actuators towards the sun. This freely-pivotable solution should be preferred to a fixed solar cell mounting on the outer hull, which is already foreseen for a covering with a circumferential aramid fabric for a meteorite protection.

If a photovoltaic system on the spaceship would be in operation, the recoverable amounts of energy would be – when placed in a distant region of the solar system – comparable to the performance of the space probe Juno, which orbited around planet Jupiter and generated there the electrical power of 490 watts by using a solar cell surface of about 45 square meters (Grammier 2009). However, considering the possible mission scenarios of the spaceship concept described here, the solar cell technology is with these results only partially or not usable at all. The achieved energy outputs are far too low to serve as a main energy supply for the engines of a multi-ton spaceship. Further arguments against the use of the solar cell technology arise also due to

- a low or omitted power generation in far distant regions caused by a reduced light intensity,

- a latent malfunction susceptibility of components to the control and alignment of the solar panels, which can be reached only with difficulty for a repair,

- an aging or damaging of the solar cells by cosmic rays that reduce the permanently recoverable energy performance.

Because the sunlight absorbed by the spaceship has generally a too low energy for a space propulsion, no further solar-based methods of energy generation have been considered either. Such a method would be, for example, the usage of the thermal difference between sunlit and shadowed hull surfaces that cannot be realized due to the therewith caused mechanical construction tensions. Also not taken into account was the installation of parabolic foil mirrors in front of the spaceship, which could bundle the arriving sunlight for a thermal energy generation at a fixed point on the hull surface.

4.2.2 Radioisotope Generator

Another way of energy production is the use of a radioisotope generator, which was already mentioned in relation to the heating of a water-based radiation protection. This type of energy generation uses a low radiating radioactive material that heats itself by its nuclear decay. Since these materials on the one hand represent a highly compact source of energy and on the other hand enable with their great half-life period a very long energy supply, which furthermore is independent from an incoming solar radiation, such generators are an ideal energy source for missions to further distant regions of space.

In order to check these units for their ability to provide the energy for a space propulsion engine, the energetic reference performances of past missions can be extrapolated. As a prominent example, the radioisotope generator of the Voyager space probes should be used here as a basis (Bennett 2006), which were launched in 1977 and that, even today, are still supplied with electrical energy. At the beginning of the mission, the overall electrical performance of such a generator was at 474 watts for which it aggregated the power of three individual generators with about 158 watts each. Their electricity generation uses the temperature difference between the introduced amount of decaying plutonium-238 dioxide $^{238}PuO_2$ and the cold of space, whereto numerous Seebeck-elements have been placed inside the generator wall that transfer the outgoing heat flow to external cooling surfaces and thereby generate electrical energy out of the Seebeck-effect.

Since the used plutonium-238 dioxide has a half-life period of 87.7 years during which its continuing decay gradually reduces the emitted heat energy, also the Voyager probes experienced a decline in their available energy output by about 1.22 percent per year. In 2015, each of the probes could be operated therefore with only 255 watts of remaining electrical power (NASA JPL 2015). When assuming for the concept spaceship a maximum usage time of 60 years and estimating an electric power requirement for the four space engines of averagely 8,000 watts, then this energy amount has to be increased by the over this period occurring decay losses to calculate the required initial generator power. But because the highest power demand occurs for long-term missions usually to mission start, and the space engines should require much less power after reaching the travel speed, only 50 percent of total 83 percent energy loss over the total usage time have been considered within the calculation as loss compensation. The so determined initial power of 12,000 watts should thus fall after 36 years for the first time under the average space engine demand. It can be assumed that in the view of this long time period, the foreseen travel speed should be already achieved, and that the remaining portion of energy can be used then only for the orientation controlling.

The three Voyager radioisotope generators use together 13.5 kilograms of plutonium-238 dioxide to provide the mentioned performance. If this mass is now up-scaled for an initial capacity of 12,000 watts, the amount of 341.8 kilograms plutonium-238 dioxide would be required for that, which would – considering a specific weight of 19.816 g/cm³ – correspond to a sphere with a diameter of 32 centimeters. Such a mass of decaying material would be also required with radioisotope generators of the newest generation, which achieve almost the same efficiency rate of 6.7 percent (Pustovalov 2007) and that require 300 grams of plutonium-238 dioxide to generate 10 watts of electrical power.

When integrating a radioisotope generator, it is important not to exceed the critical mass of the plutonium isotope. Plutonium-238 dioxide reaches its critical threshold above a mass of about ten kilograms (Blanchard *et al.* 1999). This means for the generator design that the 341.8 kilograms of isotope have to be split to at least 35 or more individual generators, of which each must be provided with an own radiation-absorbing housing. In the case of the Voyager probes, the plutonium-238 dioxide was surrounded for radiation attenuation with three beryllium capsules that together weighed 113.07 kilograms. Such a configuration is also usable for the radioisotope generators described here, whereto the housing form for the integration into the concept spaceship can be designed in a way that the least possible amount of the heavy beryllium metal is required. Another possibility to save shielding mass is given by the mounting of the generators outside the spaceship hull, which already can take over a part of the required shielding with its integrated radiation protection. This ideally should be done at the lower deck surface and close to the axis of rotation in order to reduce, on the one hand, the imbalance for the rotation flight caused by the generator mass and, on the other hand, to distribute the generator weight in landed state uniformly to all landing gears. If the radioisotope generators of the Voyager missions would be used for the intended energy supply, then there would be 26 of these units necessary, whereby the spaceship mass would be increased by a total of about 3,291 kilograms.

For cooling of the used plutonium-238 dioxide, ideally, the thermal compensation flow of the spaceship hull should be used. For this reason, it will be led near the rotation axis to the generators, where it absorbs thermal energy that afterwards can be spread over the whole spaceship hull. If the compensation flow is done with air, a part of the external cooling surfaces will still remain in operation, since a pure air based cooling does not provide a sufficient heat removal. However, once the generators can be cooled by an optionally inserted hull cavity water, it is possible to disable the external cooling surfaces completely as the circulating liquid removes already all thermal energy from the generators. For the conception of a long-term mission, these two cooling scenarios are the first choice according to the current state of the art.

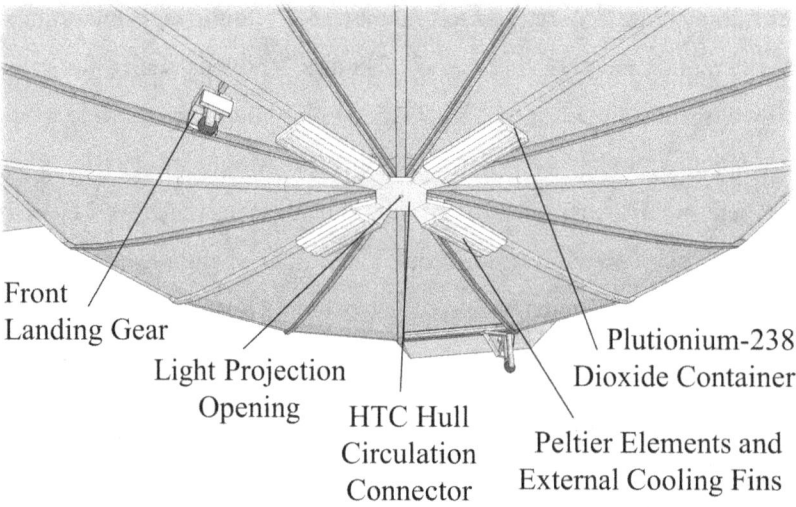

Front Landing Gear
Light Projection Opening
HTC Hull Circulation Connector
Plutionium-238 Dioxide Container
Peltier Elements and External Cooling Fins

Fig. 22: Integration of the radioisotope generators

In summary, radioisotope generators thus have the potential to ensure the heat supply of the spaceship inner for a long time and, in parallel, also to generate a stable predefined electrical energy output. The principle used for the power generation is maintenance-free whereby malfunctions of the power supply are nearly eliminated. The required additional mass for the radioisotope generators is in comparison to chemical energy sources relatively low. Furthermore, the self-heating isotope provides the option to use its thermal energy directly for a steam generation. In this case, the plutonium-238 dioxide would be surrounded by water that would be quickly evaporated by the intense heat. The thereof resulting steam pressure then could be led out from the generators and be converted directly with turbines into kinetic energy.

However, since the complexity of such a mechanical power system increases greatly, and thereby also the risk of abrasion-related failures will be enlarged during a long-term operation, its usage is recommended only for temporary used spacecraft functions. A big advantage of the mechanical steam energy is, on the other hand, the strong reduction of conversion losses, since the conversion efficiency for this type of energy transformation is at about 20 to 30 percent. If the space engines would be operated only mechanically by a steam generation, the performance equivalent of 12,000 watts, thus, could be reached already with a quarter of the aforementioned amount of plutonium-238 dioxide. In the case the entire isotope mass for the intended electrical power generation could be used, the available engine power would then even rise to 35,820 watts whereby already at mission start a far greater amount of energy would be available.

Looking now at the energy performance of decay-based energy production, this can be used very well in conjunction with the concept spaceship, if

- if the necessary flight maneuvers can be made with an electric engine power of maximum 12,000 watts, whereby each of the four intended engines would only have 3,000 watts available, which corresponds to the operation of a medium industrial motor, or

- the engines can be operated alternatively by a generated steam pressure, but which requires a mechanical conversion into motion energy, and

- the needed amount of plutonium-238 dioxide would be available.

Since this metal is very expensive due to its complex extraction and is moreover not available for private organizations because of the general regulations, a non-governmental usage can be excluded for the moment.

4.2.3 Cold Fusion Reactor

The efficient energy production from nuclear fission seems to be a perfect solution for the field of extended space travel. However, if even more energy should be provided for a space engine, but without considering the mass of a complete nuclear power plant just for this, we come inevitably to the supreme discipline of atomic energy generation: the nuclear fusion.

Therefore, with the last energy source, a theoretical alternative shall be described that although is based again on a carried fuel, which, however, can be used with such high efficiency that even a small amount of it is sufficient to provide a high energy output.

In contrast to nuclear fusion, the atomic processes in today's nuclear fission reactors are downright unspectacular. Heavy atomic nuclei are split by the impact of external neutrons and, thus, produce energy and further free neutrons. This principle continues as a chain reaction in the used fissile materials, as for example uranium-235, and splits their atomic nuclei without any additional external energy input whereby very high temperatures can already be achieved with a relatively small isotope mass. In the context of a use in terrestrial power plants, this technique also represents a carbon-neutral source of energy. It further has by the variable insertion of neutron-braking moderators the advantage of a controllability, which – in comparison to the pure decay based heat generation – can arbitrarily accelerate or suppress the fission process, which is why the energy stored in the fuel elements can be better preserved due to a demand based energy extraction (El-Genk *et al.* 2004). On the other hand, the use of nuclear fission of course includes the known risks out of a release of ionizing radiation, of nuclear accidents as well as of the problematic final storage of all formed decay products. In relation to a low-earth orbit spacecraft, also the risk from a possible nuclear contamination of the Earth's atmosphere is added to an accident scenario. And furthermore, the shielding to attenuate the high radiation of this technique has to be implemented much more massive, which leads the so increased spacecraft mass quickly to the uneconomic weight range of a chemical drive.

For this reason, with the nuclear fusion, an energy source will be described that can be largely used without the aforementioned disadvantages. If the technical state of research would allow it, the use of such a high-energetic and nuclear based energy source would be the perfect solution for spacecraft propulsion, which, on the one hand, overcomes the problem of a limited engine power and, on the other hand, makes it possible to keep the space engines permanently in operation. Therefore, the power dimensioning of a nuclear fusion reactor should be able to provide the energy for a launch from the Earth's surface, but in the following also to maintain this energy level for a long time in order to ensure the further acceleration of the spaceship.

The fusion process considered for energy production corresponds to the first stage of a solar fusion sequence. Hereby, the atomic nuclei from hydrogen atoms are fused by a so-called proton-proton reaction into the next-complex atom core structure, the deuterium D_2, which normally is based on the high temperatures in a sun. The resulting deuterium is – as well as its hydrogen basis – harmless in terms of toxicity and radiation levels. Nevertheless, a gamma radiation is emitted from the fusion reaction, since freed positrons and free electrons are completely transformed into energy by their annihilation. This radiation consequently have to be attenuated by a suitable shielding. Due to this reason, and moreover to ease the build of a fusion-based energy generation, all further fusion stages of the solar energy generation were ignored. In a fully processed fusion chain, the deuterium could fuse into helium at a higher temperature and with the emitation of additional radiation, which, however, would require a much stronger shielding.

The energy surplus, resulting from a hydrogen nucleus fusion is 0.42 MeV, which can be supplemented by further 1.022 MeV out of the following annihilation (Miramonti 2010). If we assume $6.022 \cdot 10^{22}$ potential hydrogen atoms, which are – according to Avogadro – contained in 18.015 grams of water, the maximum energy surplus from the fusion of one gram of water H_2O to deuterium oxide D_2O (heavy water) is:

$$E = \frac{6.022 \cdot 10^{22} H_2 \cdot (0.42\, MeV_{Fus} + 1.022\, MeV_{Annih.})}{18.015} MeV/g$$

$$E = 4{,}820 \cdot 10^{21} MeV/g$$

The resulting mega-electron volts can be converted also into Joule for comparison purposes. For the calculation of a real energy yield, it further has to be considered that in a realistic reaction process only a part of the hydrogen atoms will be fused. By using an assumed fusion rate of five percent, the energy yield out of one gram water thus is:

$$E = \frac{(4.820 \cdot 10^{21}\,MeV/g) \cdot 5\%}{6.241.509.647.120 MeV/J} J/g$$

$$E = \sim 38.612.453 J/g$$

If we use this energy yield values for a dimensioning calculation in which a simplified fusion reactor shall produce exemplarily 40,000 watts of thermal energy, this would process an hourly water amount of

$$m_h = \frac{40.000\,W \cdot 3.600 J/Wh}{38.612.453 J/g} g/h$$

$$m_h = \sim 3{,}73\,g/h$$

Thus, each day, a water demand of about 89.52 grams would arise that – assuming a fusion water storage of 500 liters – would ensure the continuous operation of the space engines for 15.3 years.

Since the actual fusion area of a micro fusion reactor should be less than one cubic centimeter, its construction can be located inside the spaceship hull as well. This small reactor volume further makes it possible to surround the fusion area with a sufficient lead shielding without causing an excessive weight increase thereby. Lead metal is known to be very heavy, but it however is also able to attenuate a gamma radiation of one MeV within one centimeter shielding thickness by about 50 percent (McAlister 2012). The expected gamma radiation level of 1.022 MeV by the annihilation of positrons and electrons thus requires a shield thickness of at least two centimeters whereby the radiation passage would be exponentially reduced to less than one percent of the original value. Inside the shield, there are furthermore a thermally stable enclosure of the fusion area as well as the necessary lines for the heat dissipation so as not to heat the surrounding shielding lead over its melting temperature of 327.43 °C.

The so produced fusion heat can – analogue to the decay-based energy production – be used for a steam generation. The thereof resulting steam pressure supplies the individual engines with mechanical or thermal energy and would be distributed via pressure pipes from the navigation room to the four engine gondolas. There, also a condensation of the steam can take place inside externally mounted cooling radiators, whereupon the steam cycle would be closed with the return of the condensed water into the fusion reactor.

In current large-scale experiments, it is currently assumed that for a continuous plasma fusion of, for example, deuterium ^2H and tritium ^3H – both also isotopes of hydrogen, permanently temperatures of 100 million °C must prevail. The nuclei are thereby thermally freed from their electrons and provided with sufficient kinetic energy for overcoming the Coulomb barrier, which results from the mutual repulsion of the positively charged protons. Because only if the atomic nuclei have approached each other sufficiently against the repulsive forces, so that their gravitational attraction finally exceeds these, a fusion of those can occur (Hamacher *et al.* 2001). To start such a hot nuclear fusion, thus, large amounts of energy must already have been provided before. Since there are, in addition to the thermal energy, also further energy efforts necessary for a magnetic enclosure of the superheated plasma, the energy balance of the hot nuclear fusion quickly becomes negative. Furthermore, the difficult to renew fusion plasmas are undergoing a steady contamination with eroded components out of the reactor wall by which the conditions for a self-sustaining fusion process continuously deteriorate.

A low tempered nuclear fusion can avoid these problems since the provided atomic nuclei have to be less heated. As result thereof, the fusion area can be surrounded with a simple heat-resistant reactor wall, and a less energy input is required to initiate the fusion process. Moreover, a low operating temperature allows an easy and thermally adjusted replacement of the reactor contents whereby the ongoing fusion process will be less affected.

In order to let the fusion reaction take place at a reduced thermal level, some preparatory measures are required, for which two new approaches shall be described in this concept.

For the first variant of a cold proton-proton reaction according to Fig. 23, initially the H_2O used as fuel has to be decomposed into its single atoms. This already occurs under a quite moderate temperature in the fusion area, which can dissociate the supplied water steam at 1,700 °C into its components. As a result of this process step, the fusion chamber initially also contains the free oxygen atoms of the water, which can be used after the hydrogen fusion to oxidize the produced deuterium gas during its cooling to deuterium D_2O – better known as heavy water. Thereby, on the one hand, a part of the dissociation energy will be recovered, and, on the other hand, the reactor products can be collected in a not hazardous respectively liquid form that may be easily released for mass reduction into the free space. But first it is necessary to reach and maintain the starting temperature of 1,700 °C in the foreseen reaction area, which for a specific point can be achieved, for example, by the use of a laser and correspondingly suitable ceramic materials.

After the thermal water decomposition, the hydrogen molecules respectively their atomic nuclei still have their full electron shell, which in the second step has to be removed via an ionization. The easiest method to manipulate electron shells is their bombardment with ionizing radiation. As an example, short-wave ultraviolet radiation, or X-rays below a wavelength of 50 nanometers can be named here, which are able to remove electrons from the atomic shells of hydrogen. Thereto – also with the support of a laser –

corresponding results have already been achieved (Sakabe *et al.* 2004). For this process, the ionization of one hydrogen atom requires a radiation energy of at least 13.6 eV (Speight *et al.* 2005). However, since the removal of electrons from the hydrogen atoms still does not trigger the final fusion process, but only converts the atomic nuclei into positively charged protons, the ionization is also just a preparation for further fusion-relevant measures.

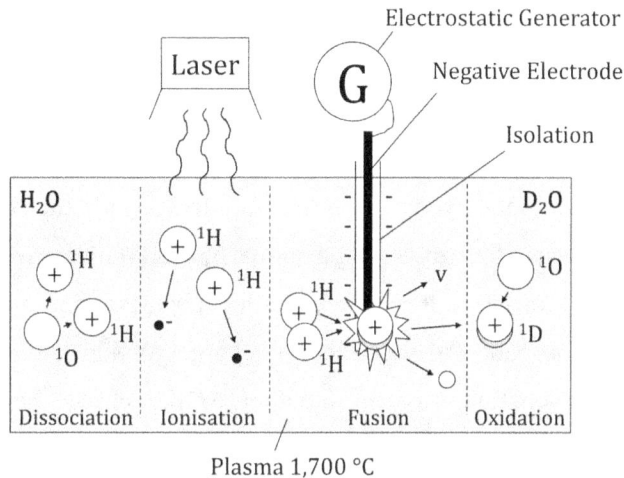

Fig. 23: *Principle of the electrostatic nuclear fusion*

The second variant to prepare a cold hydrogen fusion deviates thermally even further from the current experimental reactors and uses according to Fig. 24 a simple water-based acid as initial basis for a cold fusion. During the aqueous acid formation, hydrogen protons are dissolved out of the surrounding water molecules so that these remain in form of OH$^-$ hydroxide ions within the solution. The freed protons, instead, bind each with another molecule of water and form thereby oxonium ions H$_3$O$^+$. These ions thus hold a surplus, positively charged atomic nucleus, for which reason they are likewise a potential basis for a fusion (Toivola *et al.* 2009).

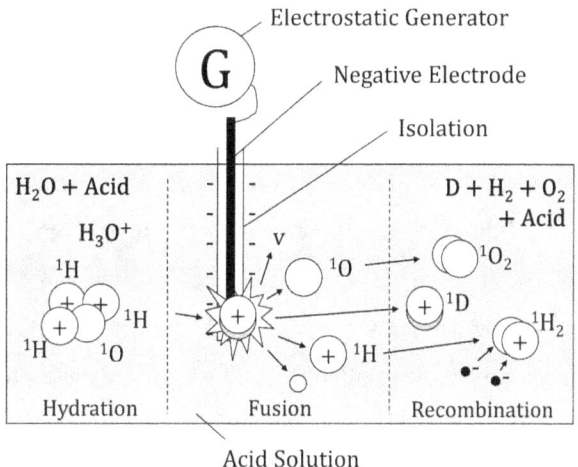

Fig. 24: *Principle of an electrostatic acid fusion*

The central objective of simplified cold nuclear fusion is now to reach a proton-proton reaction with the positively charged hydrogen protons from the two preparation variants. Thereto, the following hypothesis is set up for both variants: besides the known thermo-mechanical forces to reach a nuclear fusion, also the electrostatic compression of ionized hydrogen atoms can trigger a cold nuclear fusion. According to this, the cold nuclear fusion is based on two core factors:

- An increased density of the hydrogen protons: since a massive reactor body can very well withstand an inner pressure due to the lower temperatures, the hydrogen protons will be brought relatively close to each other.

- A catalytic effect that moves the hydrogen protons rapidly towards each other: if this effect, for example, consists of a very strong negative charge, the positive protons of the hydrogen would be attracted to this, whereby a local compression of free atomic nuclei would be formed around the source of the charge. Within this compression area, the protons push continuously in direction to the charge source, while additional accelerated protons penetrate from the outside. Thus, in dependency to the negative charge strength, the probability of overcoming the Coulomb barrier rises, which is why the fusion process may be initiated or suppressed via the charge source.

The ionization and pressure-based compression of the hydrogen atoms to be fused is technically already possible without any major efforts. Therefore, the described electrostatic compression of the atomic nuclei is as much more important, since even under the extremely hot conditions of today's test facilities, the probability for a fusion of two atomic nuclei is very low. One of the main reasons for this is, of course, the electrostatic repulsion, and it is therefore a particularity in nuclear fusion research to harness these electrostatic forces now to increase this fusion rate. Even a slight improvement of the currently achieved nuclear fusion rate would be a fundamental step forward, but here this principle shall be initially applied to the cold nuclear fusion.

For the attraction of the positive charged hydrogen protons, a statically negative charged cathode shall serve, which can be used as proton attractor in both variants of hydrogen ionization. This will be inserted directly into the center of the fusion reactor and attracts the therein present protons close to its electrode surface. If the static field will be established with appropriate strength, then – together with the thereof resulting increase of compression – also the probability of a collision of these protons rises by which their fusion would be initiated.

The cathode must be inserted completely electrically isolated into the fusion-area, since an inflow of external electrons would revoke the ionization of the protons. In this detail, especially the acid-polarized nuclear fusion differs from an electrolysis in which ordinary hydrogen gas will be formed at the negatively charged cathode, where the additional hydrogen protons are pulled out of their oxonium ions H_3O^+ by the supplied electrons and are directly bound in form of hydrogen molecules.

Inside the concept spaceship, the outer ring area of segment 5 has been foreseen for such a fusion reactor. According to the above requirements, its construction consists essentially of a pressure- and heat-resistant reaction container that is surrounded by a water-based heat dissipation and a radiation shielding made of lead. Due to the high fusion-energy gain, the reactor can be designed very small and should contain a maximum fusion chamber volume of one cubic centimeter. Since the internal temperature is at a maximum of 1,700 °C or far below, no further components are required for the thermal insulation. This fusion reactor system is supplemented by a static charge generator, an optionally required laser system as well as the liquid pumps – inclusive a balance tank to a continuous water provision of the steam generation. A simple static charge generator is already integrated with the basic equipment in the spaceship. Nevertheless, it may be necessary to convert a part of the generated heat energy directly by a steam turbine- and generator combination into electrical energy whereby the operation of a more powerful laser or high-voltage generator is made possible. In contrast to these jointly located technical components, the water fuel used for the fusion, however, should be evenly distributed around the rotation axis of the spaceship in order to avoid a rotation imbalance.

4.3 External Material Collection

If a spaceship is operated for a long time without the energy intake of a high-energy source, the gradual energy losses of a biologic-process-based life support system will quickly become a survival important topic. Since a spacecraft in space represents an open thermodynamic system that can take up energy and can release it as well, the entropy contained therein moves – assuming a predominant energy output and following the second law of thermodynamics – over irreversible processes to an ever higher level (Sheehan 2011). For the concept spaceship this means figuratively, if energy is generated within the life support system and is discharged out of the spaceship to the outside, all thereto involved molecular resources would consume itself over the time so that, finally, only materials with the lowest chemical energy content would be available.

In general, such outgoing energies consist out of the radiated heat energy and a produced thrust of the space engines. A radioisotope- or fusion generator can easily compensate these losses during a long-duration mission by its released heat energy. At missions in the inner region of the solar system it is moreover possible to do a thermal stabilization by the available solar heat radiation for which additional external heat collectors or radiators can be used. But, however, it would be more difficult to compensate the steady loss of energy in a scenario in which the life support system would have to last without a high-energy source for the intended usage time of 60 years. Because in this case it becomes necessary to add constantly new usable matter to the biospheric life support system.

A possible source of matter during a space flight are the dusts and micrometeorites in space – as they daily fall to earth with an estimated mass of 82 tons and a variable chemical composition (Engrand 2011). In free space, this matter moves in diffuse flight-directions and must therefore be actively collected by a statically charged element. At segment 5, such a collection unit has been placed within the so-called material collection bay (MCB).

The MCB uses the entire upper deck surface section of segment 5 by extending laterally with its opening along the radial segment frames and ending in the center directly at the central tunnel wall. With the outer edge, it ends about 66 centimeters before the vertical outer wall. Inside the MCB extends a horizontal surface, which is – analogous to the structure of the outer walls – also provided with an inner and an outer hull planking as well as the mentioned inner-sided insulation- and radiation protection measures. Above the MCB opening, a movable closure cover is placed for an optional protection that – viewed from the outside – has been mounted with hinges on the left side. The closure cover has the same curvature as the upper deck surface, so that the aerodynamically rounded hull shape is kept as long as it is closed. The opening and closing is done opposite from the hinge-side by a simple toggle lever mechanism, which is ideally operated by a drive that is not prone to malfunctions. The required opening angle of the MCB cover during a material collection is at 45 degrees by which the centrifugal force based transport of the collected dust is eased. Consequently, the collection of cosmic dusts is also mandatorily bound to an artificial gravity that drives the transportation of the matter into the spaceship inner without further measures.

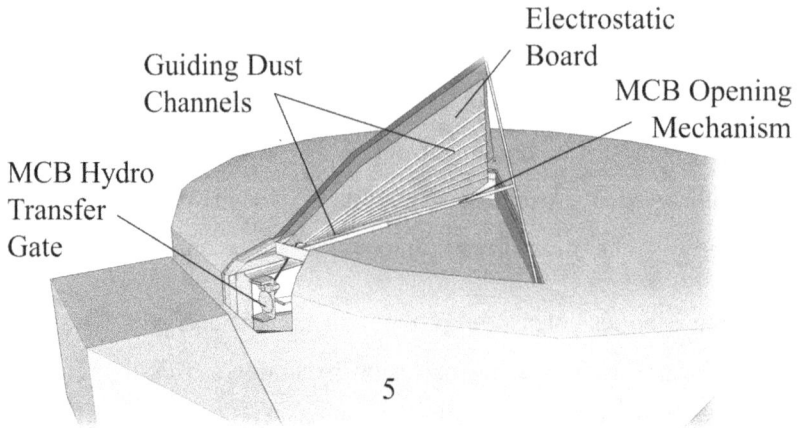

Fig. 25: View of the MCB

To attract the cosmic dusts and guide them controlled to the MCB surface, there is an electrically insulated panel made of plastic located at the inside of the closure cover. This has been attached below the closure cover hinges to the MCB surface and is movable as well. In order to open and close the panel simultaneously with the closing cover, it is connected to the closing cover on the opening side by three flexible spacers. The panel has further an electrical line to the inner of the 5th segment, where it can be connected to a high voltage generator and be electrostatically charged therefrom. The principle of the charged MCB panel thereby is equal to an electrostatic duster: small particles are first attracted and then held by an externally provided charge. But if in the following the charge is removed again, the particles can move on and would immediately be pulled by the rotational centrifugal force towards the outer side of the spaceship. In order to prevent thereby an escape of the dust from the material collection, the charged panel is provided with diagonal fins, which extend in outward direction with 45 degrees downwards to the MCB surface. Along these, the particles gradually move into the lower inner corner of the MCB where they finally fall to the outer end of the MCB. At this point, the downpipe described with the hull construction is located, which takes over the incoming matter and forwards it to the inside. The top of the downpipe is provided for this purpose with a protruding cone and a protection grid against too large objects. In the inner of segment 5, this downpipe can be looked inside via some glass lenses to determine its filling level. If a sufficient amount of dust has been collected, its manual takeover takes place from time to time by a therewith connected hydrodynamic lock.

4.3.1 Hydrodynamic Material Lock

The operation of this material lock type is done by hand and is essentially based on a rapidly rotating liquid ring, which can create an air-tight and pressure-resistant connection to the inner vacuum of the downpipe via the supplied centrifugal force. The principle described below thus can carry out a material takeover from space without the loss of valuable respiration air, which is why the lockage operations can be done more frequently without consuming the own resources. For this purpose, the hand crank of the lock drives a fast rotated lock shell that is half filled with a transfer liquid, so that the preferably heavy liquid is pressed by the centrifugal force to the radial outer side of the shell. The top of the lock shell has a small centered opening into which the material downpipe extends contact-free. The pipe ends therein with a circular plate, which is immersed during operation on all sides into the rotating liquid. Below this plate there is a closure plate of the same diameter installed that, if the lock is not used, is pressed together with the lock shell against the downpipe plate whereby the thereon applied rubber seal forms outwardly a permanent mechanical lock closure. The closure plate is rotatably mounted with a bearing onto the bottom of the lock shell, so that an acceleration of the shell can be done before the mechanical seal is opened by the backward movement of the lock shell.

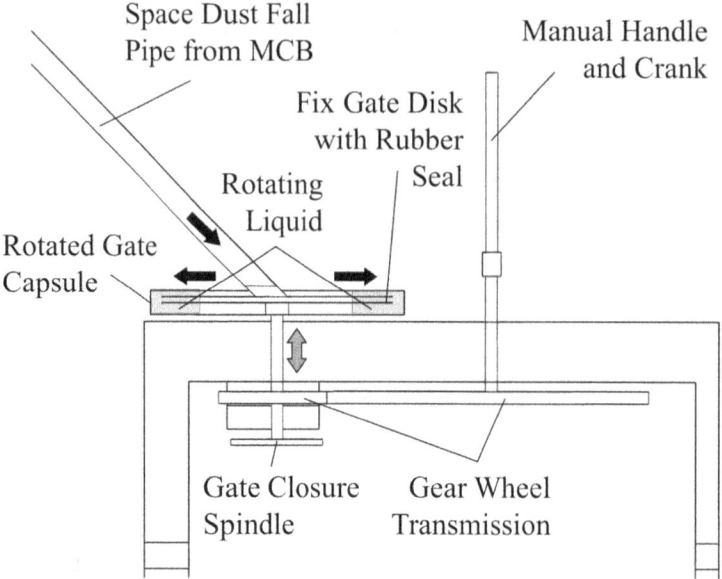

Fig. 26: Hydrodynamic material lock

If the material lock is opened, the strong centrifugal force of the rotating liquid prevents its outflow into the sucking vacuum of the downpipe. This can be compared with the suction of a fluid from a great height. If the air is sucked out from a vertical tube at its upper end, a liquid, supplied at the lower, end will be pressed into the tube by the outer air pressure. But, however, the maximum suction height is limited by the weight of the liquid as well as by the externally applied atmospheric pressure. Therefore, the weight force of the transfer liquid will be increased via the rapid acceleration in the lock shell, so that its radial rise to the central opening of the downpipe should be prevented. During the lockage operation, the dusts trickle out of the downpipe into the circulating liquid and are deposited along the inside of the rotating shell. If the lockage operation is completed, the lock shell is pressed together with the closure plate again against the downpipe plate, during which the transfer liquid cleans automatically the sealing rubber coating, whereupon the downpipe opening can be tightly sealed again.

Since the usage of the MCB is – as described – only possible during a rotation flight, the material lock consequently has been arranged for a sole use under a radial acting gravity force. The unusual hand drive for the lock shell serves, of course, to a reduction of the lock complexity, but it also suggests that the use of the lock makes only sense if, in parallel, also a planting is operated, which anyway requires the physical involvement of at least one space traveler.

4.3.2 Static Charge Generator

Another component of the material collection is the in segment 5 inserted generator for static charge (GSC). Its bearing and driving technique is based on the floating direct current motor described in the chapter on the space engines, which is already used in a similar design for transporting the nutrient solution. Since this generator optionally can also be used to operate an electrostatic fusion reactor, it must be able to function under both possible gravity directions. Therefore, its gravity-bound bearing and driving unit has been mounted to a rotatable suspension that allows the generator to follow the gravity with its lower end. The water supply of the contained hydromagnetic bearings is provided from the drinking water tank in segment 4 for which the generator has been placed close to the lower deck surface in order to allow the gravitative water inflow in a landed state as well. On the generator axis, an electrically isolated mounted plastic disk is located that is being provided with an electrostatic charge by fixed and unrotated brushes. This charge can in the following be picked up via the same brushes and can be forwarded to the MCB or a cold fusion generator.

4.4 Energy Dispatching Nets

Since the energy generation of the concept spaceship described here has been implemented with various techniques for a plurality of consumers, the summarizing balances of the thereto involved electrical and thermal energy sources (see Tables 7 and 8) can be used to ease the creation of a final energy concept. The control of all energy flows should be done centrally through the navigation room, which is located in the inner area of segment 1 and to which the energy lines of all fuel cells, generators, and reactors are initially leading. For the subsequent energy distribution, two separate energy nets are foreseen whose consumers can be supplied with energy in thematic groups via a control panel (see Table 9).

The first energy net connects the electrical basis energy of the fuel cells with the elementary core systems of the spaceship, which are the plant illumination, the air- and nutrient solution movement, the material collection as well as the on-board computer, and the navigation sensors. This net is basically designed for 24 volts direct current. In order to control the plant illumination, separate cables have to be laid into the planting segments in order to enable, if necessary, a variable control of the lighting rhythms. All other consumers will be connected via a direct cabling with the navigation room.

The second energy net manages the energies of the space engines and may – in accordance to the chosen type of engine energy – consist of an electrical as well as an optionally inserted steam-pressure-based line system. This distribution system comprises also the controlling of the four engine gondolas and their space engines as well as the provision of possibly required energy demands to operate a cold fusion reactor. Depending on the needs of this reactor, it may also be necessary to locate the complete controlling of the fusion process into the navigation room for which purpose further electrical lines must be optionally taken into consideration.

5 Simplified Apparatuses for Electro Mechanics, Navigation and Spacecraft Propulsion

5.1 Internal Drives, Environment Sensors and Navigation Solutions

The before mentioned components to the propulsion and controlling of the concept spaceship appeared so far only as potential energy consumers. The following chapter is now devoted to these units and provides more detailed information about their integration and their way of functioning.

5.1.1 Hydromagnetic Bearings, Floating Direct Current Motors and Magnetic Gearwheels

As central components of the life support system, two electric direct current motors have been integrated within the spaceship concept that perform the elementary air- and nutrient solution movement and drive a high voltage generator. Considering the hereto necessary long period of operation, a special bearing- and driving concept was chosen for these motors, which allows a completely abrasion and maintenance-free operation.

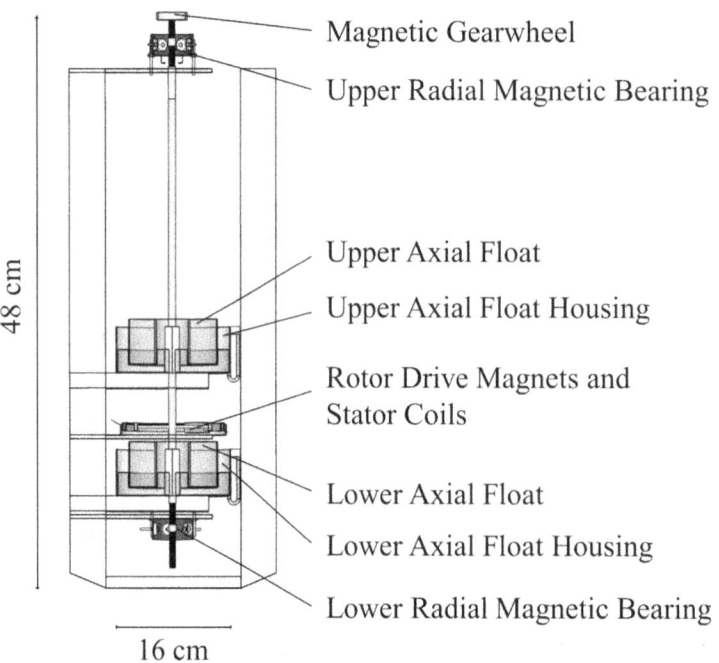

Fig. 27: Cut view of the floating direct current motor

Thereto, the floating direct current motors generate – by using very simple and durable electronic components – an alternating inner electromagnetic field that is used for a continuous drive of the central motor shaft. A particular aspect of this motor type is further the use of axis floats and radial centering magnets, which provide a contact-free axis bearing by the inclusion of only gravity. Furthermore, this avoids the use of any other electronic control components, since sole the filling level of the surrounding axis float housing regulates the magnetic interaction of all bearing components. For this reason, there is also a minimum water inflow required into the float housings, which hold with their implemented overflow barriers automatically the required liquid level and supplement so simultaneously all evaporated water portions.

Since all components of the motors can be completely encapsulated or encasted, a corrosive reaction with the used water can be excluded. So could, for example, the wound motor stator fully be casted by a synthetic resin material, and also the bearing magnets would operate non-lubricated and abrasion-free under a plastic layer. The electronic components – which consist for a single motor only of two power transistors, two foil capacitors and two resistors – were selected based on practical preliminary tests in order to ensure their potential long-life sustainability. Under optimal environmental conditions is already hereby the correspondingly long usage duration of all components enabled.

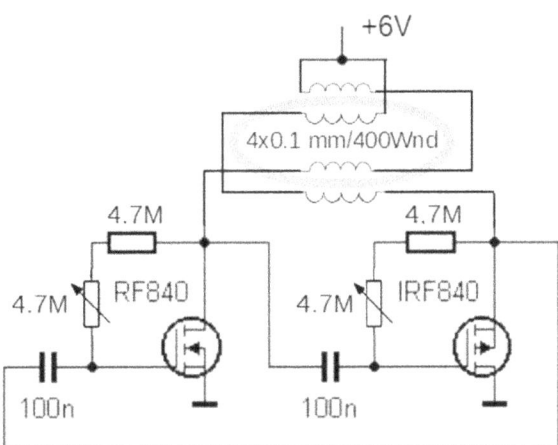

Fig. 28: Circuit of the floating direct current motor

The circuit used for the alternating electromagnetic field represents a simple solution for an alternated power supply onto each two coils of the motor stator, which from this generate an accordingly alternating magnetic field around the rotor magnet. The energy consumption of such a motor can easily be defined by the implemented coils. In terms of energy efficiency, no optimal values can be expected due to the increased bearing fluid resistance, but is for the intended motor usage a slow rotating axis be sufficient, which in addition can be further translated by a followed magnetic gear transmission.

A special variant of the hydromagnetic bearing described before has been used to the suspension of the nutrient solution screw pump in segment 1 (see Fig. 10). The particular feature is hereby the bearing of the axis having an inclination angle of 45 degrees. For this purpose, strong bar magnets are located inside the ends of the pump axis by which they are provided with a magnetic pole. Directly in front of the outer end of the axis, several stator magnets are located at the lower pump float housing. These are counteracting against the axis pole and that absorb all horizontal forces resulting out of the inclined position, while the floating body of the pump axis carries the vertical weight forces. Moreover, lateral magnets below the pump axis provide a further centering.

The upper end of the pump, which extends through the ring corridor separation wall to the nutrient solution distribution channel, requires a supporting bearing as well. Here, four additional magnets are used to lift and center the pump axis via its repulsive pole.

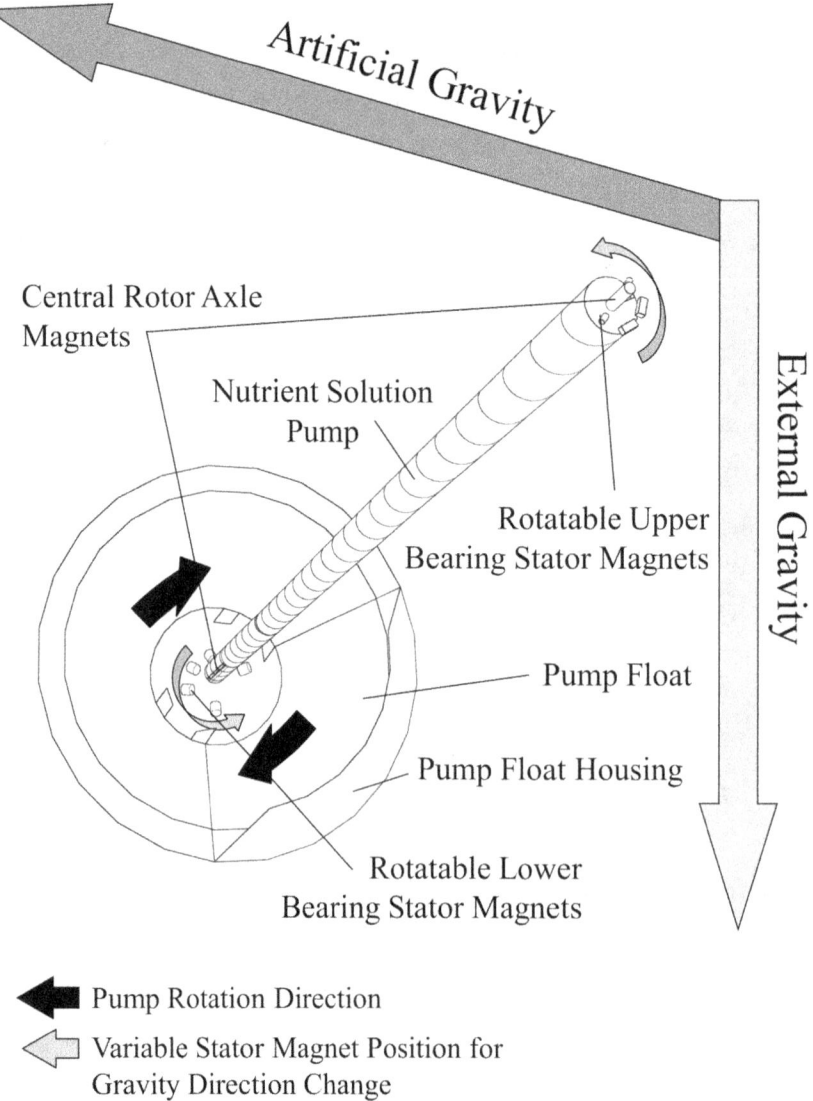

Fig. 29: Cut view of the nutrient solution pump bearing

In order to reconfigure the hydromagnetic bearing at a gravity change, all magnet combinations of the bearing stator are rotatably mounted towards the pump axis. By the rotation of these stator magnets by each 180 degrees, the bearing operation can be enabled also under a 90 degrees changed gravity direction.

In conjunction with this abrasion-free axis bearing, also a maintenance-free and fail-safe solution should be preferred for the gear-translations. Moreover, even the simplest ordinary gearwheels require a permanent lubrication in order to ensure their smooth operation and to prevent a mechanical abrasion. And because the production of such a lubricant cannot be ensured by the intended life support system, an appropriate alternative has been developed in the scope of this spaceship concept.

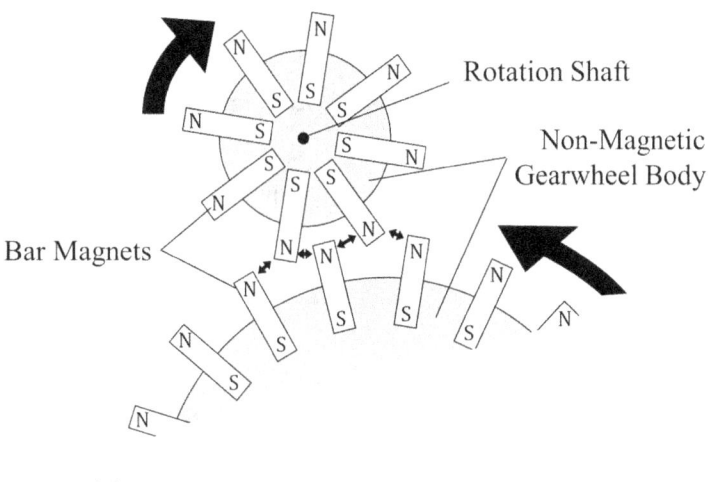

Fig. 30: Functional scheme of a magnetic gear

A contact-free magnetic gear transmission uses therefore the repulsive forces of radially mounted permanent magnets to transfer a rotational force between parallel or with an angle arranged axes. Analogous to conventional gearwheels, it is possible to vary the diameter of such a magnetic gearwheel as well as its number of mounted teeth magnets whereby an increasing or decreasing transmission can be achieved. During the experiments on this subject, it was already possible to verify that such a power transmission works also under the conditions of a hydromagnetical bearing, whereby the operability of the foreseen gear transmission in segment 1 between a floating direct current motor and the nutrient solution pump has been confirmed.

5.1.2 Navigation Projection System

In addition to the mechanical components, the navigation and communication instruments are also among the elementary components of the concept spaceship. Since, for safety and stability reasons, no greater-sized windows have been inserted into the spaceship hull, an optical visualization system is of the highest significance.

In contrast to the current use of electronic orientation- and navigation systems in space travel, with having an assumed mission duration of 60 years, no such complex cameras and monitors may be used. Instead, the required external views will be projected via lenses and mirrors within segment 1 onto a projection screen of the navigation room. The therein located display- and control unit CINA (central installation for navigation and automated processes) has been especially designed to the accommodation of all energy controls and sensor displays, which is why the projection screen for navigation is located there as well.

For the entry of optical light, only three transparent glass sticks have been lead through the spaceship hull at very exclusive positions. These sticks have a length of about 30 centimeters, a diameter of about five centimeters, and are made of either a massive optical glass or an equivalent acrylic glass. Their end faces are formed so that they do not refract the light beams, but transfer them as a straight beam into the spaceship inner.

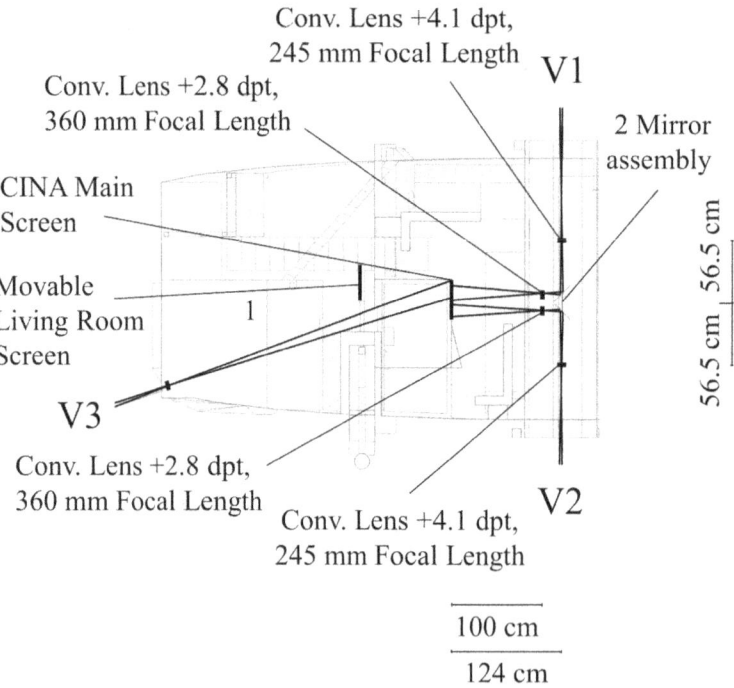

Fig. 31: Navigation projections on the CINA

Two of these sticks are located along the rotation axis and pass through the upper and lower frame structure each exactly in the center of the central tunnel. To focus the incoming light, a bundling lens has been inserted into the beam path of each glass stick, whose vertical position is aligned relatively to the middle height of the projection screen. Above and below the so defined projection level, these lenses are mounted in a distance of each 56.5 centimeters. Their refractive power is 4.1 diopters, and they have a focal length of 245 millimeters.

Directly at the height of the projection level, the from above and below arriving light beams are deflected into segment 1 by two mirrors on the top of each other, which have an inclination of 45 degrees. The upper light beam thereby is directed statically adjusted to an upper right panel of the projection screen, while the bottom mirror projects its external view on a separate projection field directly below. For a sharp projection, each an additional convex lens is required after the deflection mirrors with a light refraction of 2.8 diopter and a focal length of 360 millimeters. These projection lenses are mounted with a distance of 24 centimeters to the in mirror center crossing rotation axis and within the inner area of the central tunnel as well.

A further external view is provided by the third optical glass stick that in segment 1 has been obliquely led through the outer-edge of the lower deck surface. This position allows a view downwards from the front, which enables the navigation during horizontal flights along the surface of an astronomical object. Due to the downward inclination of 18 degrees, the front view can be displayed to the left side of the CINA projection screen to which only an adjusted lens arrangement analogue to the rotation axes projections is required. In order to allow the external front projection to the CINA, their adjoining ring corridor separation wall additionally has been provided with an appropriate opening. On the side of the living room, this can be covered by a sliding, second projection screen so that the external front view is visible on demand from the living area.

The projection screen of the CINA is aligned with its 50 centimeters high surface vertically to the spaceship front pyramid tip and is located exactly 124 centimeters in front of the central rotation axis. The profile construction required for its fixation has a height of 98.6 centimeters at this distance and is attached to the lower deck surface and the ring corridor separation wall. In order to provide the CINA with a sufficient surface for the arrangement of all further instruments, it has at its front edge a width of 68.5 centimeters, of which only 48.5 centimeters are used for the projection screen.

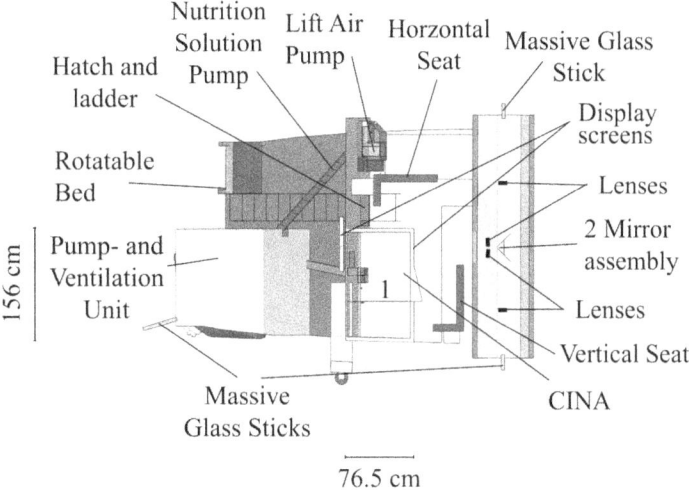

Fig. 32: Arrangement of the navigation room

The navigation room is equipped with two pilot seats to allow the use of the CINA during both intended gravity situations. A first seat is attached to the lower deck surface between the CINA and the central tunnel. It is located quite close to the deck surface so that the height of the seat and the pilot therein do not extend into the projection opening of the central tunnel. The second seat has been mounted horizontally on the ring corridor separation wall and uses the space between the CINA and upper deck surface. This seating position is ideal for the rotation flight, during which the projection screens of the CINA as well as their control instruments can be used having a table-like position.

At the further equipment of the navigation room, it was taken care to preserve the freedom of movement under both gravity directions as much as possible, which is why fixed walkable areas of the ring corridor separation wall and the lower deck surface were deliberately kept free from installations. From this results the CINA as an ideal location for the navigation and control technic, in which only the light inlet channel of the external front projection has to be considered.

5.1.3 Environment Sensors and Control Instruments

The sensors and instruments used for the CINA should be based preferably on mechanical or electromechanical constructions. Examples thereto are liquid-in-glass thermometers, barometers with simple analog measured value encoders, or electromechanical moving-coil instruments. Such instruments do not require miniaturized electronic circuits and can possibly be repaired with the given resources, which is why in principle they can be considered as long-term usable.

Temperature sensors of that kind are intended for measurements within the internal air circulation as well as for the steam inlet to the water condensers. In addition, also the temperatures of the fermentation shells and of the electrolyte solution around the fuel cells will be captured. The measurement of the outer hull temperature occurs in dependency to the realized variant for the hull temperature compensation either from the circulating air- or water stream. In the field of high energy, an additional temperature measurement will be required according to the used energy source that can be a radioisotope generator or a cold nuclear fusion reactor. Here, in particular the medium used for cooling or a generated working steam has to be permanently monitored for reasons of safety. Furthermore, requires a thereto connected high-energy steam generation the measurement of the available working pressure.

Further pressure sensors are necessary in the area of internal energy and life support, for which a continuous measurement of the inner hull pressure and the relative thereto generated gas pressure from the biogas production is done. If required, also the external air pressure is measured via an external pressure line at the front landing gear housing. In addition, the pressure compensation valve for the Earth ascent flight is connected to this line as well, which controls the required air outlet by an electromechanical opening mechanism. In combination with the internal air pressure sensors, therewith the hull pressure can be automatically lowered to the intended 0.8 bar. If then the planned inner pressure has been reached, the electromagnetic pressure compensation valve is secured additionally by an upstream manual valve. Finally, in the scope of the environment sensor measurements, also the filling levels of the drinking water tanks, of the nitrification- and balance water tanks as well as of the fuel cell electrolyte shells are detected by simple resistance-based sensors with a coarse resolution.

In order to control the outer lying space engines, there are external sensors required on the outside of the engine segments as well. First, spring metal based contact switches have to be foreseen to detect the neutral position of the engine gondolas (PU 1 to 4). In addition are switches of this type used in combination with a rotated mechanical sensor disk, whereby a very precise engine positioning can be done based on generated electrical impulses. Thereto, the gondolas should be moved either by electromagnetic actuators or by pneumatically operated air cylinders.

As a further kind of navigation instrument, also a switchable external lighting can be seen. Since the closed spaceship hull design already defines three predetermined external views, the ideal illumination area is directly ahead and behind the front landing gear, as in general only these zones can be viewed via the projection system. The light in this area may be particularly helpful during takeoff and landing operations, which is why such light units are recommended at the suspension of the front landing gear

Supplementary to these explanations provide the Tables 9a, 9b and 9c a complete list of all CINA instruments, which summarize them separated by function groups.

Table 9a: Sensors and controls for flight operations

- Display of the rotation and inclination angles of the space engines 1-4
- Operating lever to adjust the rotation and inclination angle of the space engines 1-4
- Thrust control for the engines 1-4
- Artificial horizon for horizontal flight
- Magnetic compass
- Switch for the light-controlled rotation flight TABAS
- Switch for the light-controlled space engine control LCP
- Filling level of the denitrification and balancing tanks DBT 1-4
- Operating lever of the landing gear steering
- Operating lever for the pressure cylinders to retract and extend the landing gear
- Switch for the landing gear illumination

Table 9b: Sensors and controls for the environment

- Main switch of the plant illumination
- Switch for the engine of the nutrient solution pump and the air ventilation
- Switch for the navigation room illumination
- Switch for the inner ring corridor illumination
- Display of the gravitative downward gravity
- Display of the radial rotation gravity
- Internal pressure display
- External pressure display
- Manual external valve
- Outside temperature display
- Temperature display of the hull temperature circulation HTC
- Temperature display of the internal air circulation ISA
- Steam temperature of the drinking water condensers 1-4
- Filling level indicators of the drinking water tanks CWT 1-4
- Operating lever for moving the living room projection screen
- Spaceship chronometer
- Screen of the CLEO computer
- Cable connected console for the CLEO computer
- Radio communication system

Table 9c: Sensors and controls of the power supply

Internal power supply IBE:
- Temperature inside the fermenter shells of the bioreactors 1+2
- Operating pressure of the biogas
- Biogas filling level of the biogas tanks 1+2
- Temperature of the methane gas reformers 1+2
- Voltage of the fuel cell blocks 1+2
- Current of the fuel cell blocks 1+2
- Filling level of electrolytic solution of the fuel cell blocks 1+2
- Electrolyte solution temperature of the fuel cell blocks 1+2

High energy sources RBE/FBE:
- Temperature of the radioisotope generator or fusion reactor
- Steam pressure from a radioisotope generator or fusion reactor
- Filling level of water fuel for a cold fusion reactor
- Control of fuel water supply of a cold fusion reactor

Material collection:
- Switch for the motor of the high voltage generator GSC of the MCB
- Operating lever for the MCB cover

All measurement data displayed on the CINA will be passed additionally to an input interface of the board computer in order to enable also their digital evaluation. In combination with a corresponding output interface, the automatic control of the energy-, navigation- and propulsion components is thus made possible by the CLEO operating system. For an interaction with the board computer, in addition to the central computer unit, also a variably switchable screen and a cable-based keyboard have to be integrated into the CINA. Furthermore, can the CLEO operating system be controlled by voice commands, which is why a microphone and speaker network should be integrated along the ring corridors of the spaceship. Additional details to the used CLEO operating system and thereto required components are described in the following chapter about the on-board software.

The spatial arrangement of the instruments is based on the taken pilot position for an acceleration- and atmospheric flight as well as for a rotation flight. For this purpose, Fig. 33 shows in a rough structure, which CINA instruments should be considered as useful from the perspective of each pilot seat. Following this logic, therefore, all instruments of life support and energy supply, which have to be monitored in all flight phases, are located laterally the projection screens and can be viewed relatively good out of both seating positions.

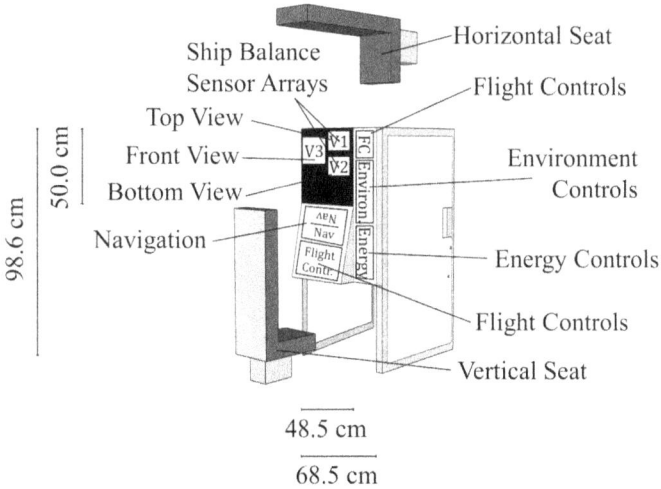

Fig. 33: Arrangement of the CINA instruments

5.1.4 Optical Display Scanning and Rotation Balancing

The concept spaceship described here relies with its optical space navigation solely on a simplified environment projection. But, nevertheless, this easily can be used for an automatic position control, which is why the CINA has been provided with a rudimentary optical scanning as well. For this purpose, the projection screens of the upper and lower rotation axes are equipped with each nine low-resistance light sensors, so-called LDRs (light dependent resistors). The used reference sensors of the type VT 93 N1 are silicon-based, and they change their resistance solely based on the intensity of the provided light. In conjunction with a star- and planet detection, however, only very low light intensities will hit these components whereby their light-dependent aging can be extremely reduced and their life duration will be extended accordingly. During the executed experiments, this type of sensor has been proven as suitable to measure the brightness of a projected planet. As preparation for these measurements, such a sensor, which has a specific dark resistivity of 300 kilo ohms, was applied with a measurement voltage of 28 volts. Because LDR sensors first need to discharge when put in darkness, a subsequent calibration time of 30 minutes was waited, during which the output voltage had been continuously measured. After this calibration-phase, an output voltage of 71 millivolts had adjusted at the sensor, which was taken as reference value to an orientation of the projection system to the dark night sky under the given stray light shielding. By the followed rotation of the projection system onto a bright celestial object, measurable changes could be determined in the output voltage of the light sensor, which are listed in Table 10 for various celestial objects in comparison to their corresponding apparent brightness. The projection system build for the experiment had widely the configuration of the concept spaceship and used the same lenses and distances.

Table 10: Sensor detection of celestial objects

Target object	Measured voltage (mV)	Apparent brightness (mag)
Dark space	71	0.00
Moon (full moon)	236	-12.73
Jupiter	99	-2.94
Venus	75	-4.67

The voltages received from this type of projection screen scanning will require an amplifier circuit to their further usage, which is implemented for each individual sensor by using a power transistor. These switching components have – due to the expected low current strengths – an extraordinary slow aging, so that in encapsulated form they can be well used for long-term missions.

In comparison to the currently in space flight used highly sensitive navigation sensors, such as, for example, the high-resolution CCD image sensors (Romanishin 2006) or photomultiplier tubes (Engstrom 1980), the performance of the here described sensor technique is inferior in many cases, but their output is nevertheless sufficient for a simple position control. For this purpose, nine brightness sensors have been inserted in a square arrangement into the center of each projection screen. If now the projected image of a star, a planet, or a moon points onto the central sensor, so can following position changes of the spaceship body be registered from the movement of the incoming light spot over the adjoining sensors. Because the projection out of the central tunnel provides permanently both views along the rotation axis, their corresponding display screens allow the simultaneous monitoring of two celestial objects as well.

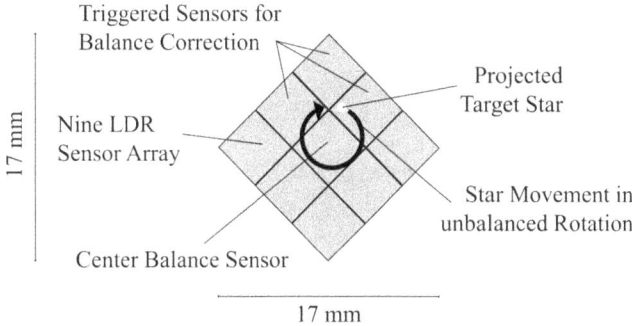

Fig. 34: Sensor arrangement for movement detection

The target object monitoring from the perspective of the spaceship center enables their usage in two different ways. If the spaceship is initially in a rotation flight and if thereby solely the central brightness sensor is lighted by a targeted celestial body, the rotation balance of the spaceship is provenly balanced. Out of this initial position, now a purposefully acceleration can be done towards to the targeted object via the uniform thrust of the four space engines. If, instead, the projected light spot cyclically leaves the central sensor, the signal of the each lighted sensors can be used to trigger automatically a predetermined balancing reaction for re-establishing the rotation balance. The integration of this rotation control is moreover possible without the involvement of a navigation computer by using only simple electrical circuits to control the ongoing adjustment of the rotation axes alignment. During a rotation flight, the tank balancing adjustment system (TABAS) is used for this, which performs the necessary mass center balancing accordingly.

As described in the chapter on life support, four denitrification and balancing tanks (DBT 1-4) are located along the outer ring corridor, which are evenly distributed around the rotation axis and that are available for the mass balancing with a total filling volume of 176 liters. For this purpose, the tanks use fall- and rising pipes, which are laid in an arc along the lower deck surface, and by their assistance, these either can be emptied or filled out of the inner nutrient solution distribution channel. Since only 50 percent of their volume is intended for the denitrification, each tank can be filled with additional 44 liters of nutrient solution. If at the same time the corresponding opposite tank is emptied, this mass would be completely shifted to one side of the spaceship disk. With the occurrence of an imbalance that puts – comparable to a spinning top – the rotation axis of the spaceship into a precession (Butikov 2006), the TABAS then changes directly the filling levels of the four balance tanks via eight simple electromagnetic valves. Because the liquid flows solely by the nutrient solution distribution and the radial artificial gravity into or out of the tanks, the use of extra pumps becomes unnecessary. Particularly rapid mass relocations, moreover, can be achieved by the spontaneous emptying of one single balance tank.

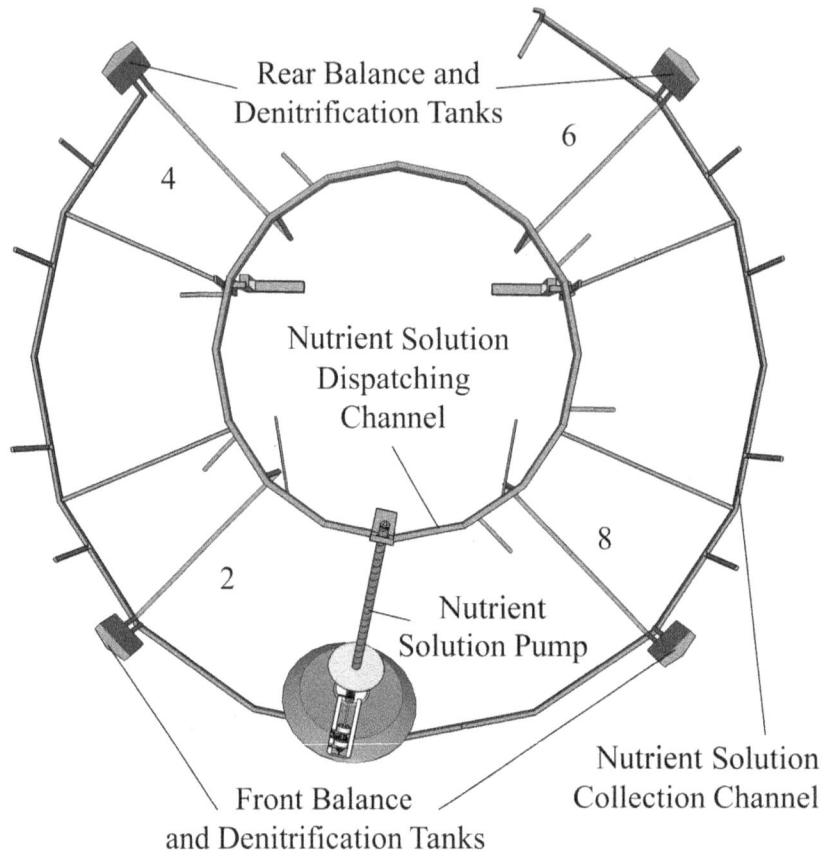

Fig. 35: Denitrification and balancing tank system

If, instead, the optical navigation system is needed for an unrotated flight phase, the spaceship orientation can be done alternatively via the four space engines by guiding the projected light spot back into the central sensor field over slightly differing thrust strengths. This functionality of a light-bound control of propulsion (LCP) is used, for example, during the acceleration or deceleration phases in order to keep automatically the flight direction vector of the spaceship.

Both navigation modes have been integrated into a TABAS / LCP computer simulation that was able to compensate a spontaneous mass imbalance or a precision movement by the aforementioned measures. The TABAS had therein an imbalance reaction threshold of 55 kilograms, with their overstepping the simulated star projection left the central brightness sensor for the first time. Having an assumed inflow and outflow rate of 1.5 liters per minute for each balance tank, a furthermore added imbalance mass of 35 kilograms have been balanced within 226 spaceship revolutions (18.50 minutes). This time span is directly connected to the available liquid flow rate, which is predetermined by the available surplus pump capacity of the nutrient solution pump. In comparison thereto, needed a precession balancing over the LCP and the space engines only 182 spaceship revolutions (15.17 minutes), whereto a precision movement at the spaceship outer side of 0.4 meters had been reduced that long that it was completely below the reaction threshold of 0.22 meters. Here, the needed time can be explained as well, since with the combination of a precession- and a rotation movement simultaneously two overlaying circular movements are displayed via the projection system, which is why the waiting time for the next correction possibility extends accordingly.

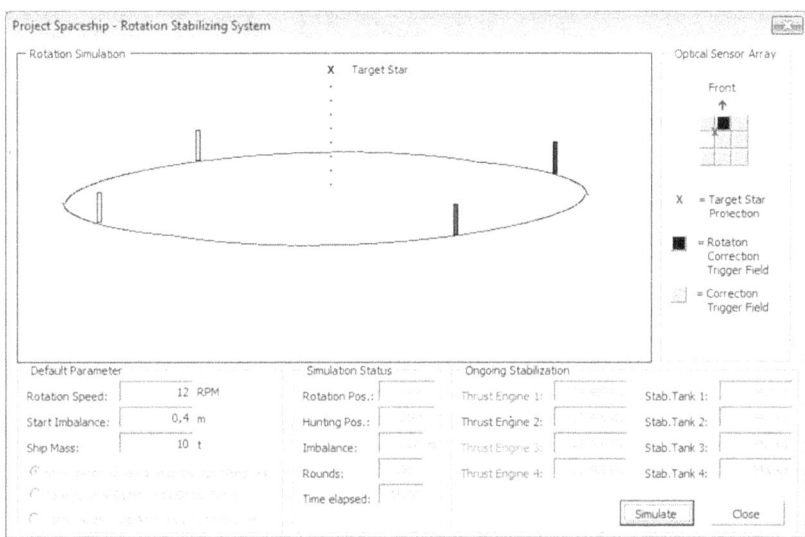

Fig. 36: TABAS/LCP computer simulation

5.1.5 Antennas and Radio Communication System

In addition to the optical view, also a radio communication system represents an important component for the connectivity to the outer environment. Since the foreseen mission scenarios of the concept spaceship require in all short-term and long-term variants a good functioning communication, this topic received an increased constructive focus. For the quality of a radio link, the size of the used transmitting- and receiving antenna is of essential importance, which is why the front pyramid with its large volume is an ideal mounting location within the spaceship construction. If therein two parabolic antennas will be hung up with their back to each other, a maximum antenna diameter of 128 centimeters can be used. This antenna size is comparable to the antenna of the space probe DAWN, which enabled at a maximum output power of 200 watts and over a distance of three astronomical units (about 449 million kilometers) a stable radio connection to Earth (Rayman *et al.* 2006).

In order to operate a similar radio transmitter, the internal electrical energy reserves of the concept spaceship are normally not sufficient. If this cannot be supported by an external high-energy source, the active plant illumination first has to be deactivated before the radio unit is put into operation. But since the photosynthesis of the plants can be interrupted only for a few minutes, the following usage duration of the communication unit must be kept very short. In practical operation, this can be achieved, for example, by using temporal predetermined transmission windows and the previous calculation of an optimal communication time, during which a transfer of compressed data packets can be assured with the best possible antenna alignment of both communication partners.

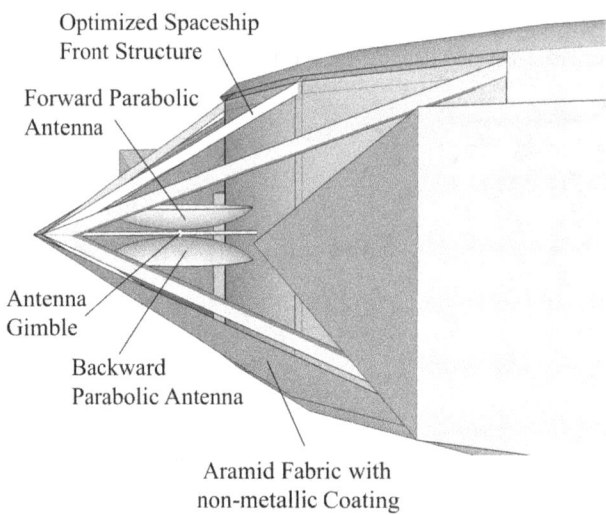

Fig. 37: Transmitting- and receiving antennas at the spaceship front

If required, can the transmitting- and receiving antennas be inclined by some degrees via electromechanical actuators or pneumatic cylinders, which enables a fine adjustment over two rotation axis. Furthermore, with the opposing antenna alignment, it is also possible to select the best suitable parabolic antenna according to the spaceship orientation. For an unhindered transmitting and receiving of the radio signals inside the front pyramid, this has been covered with only a non-metallic coated aramid fabric that protects the antennas from meteorite impacts, but still allows electromagnetic waves to pass.

5.2 Spaceship Propulsion

The choice of a suitable space engine defines decisively for what missions the concept spaceship may be used in the future. By using the high-energy sources of the chapter on energy supply, several engine variants can be taken into consideration, whose integration will be contemplated in detail in the following.

5.2.1 Space Engine Suspension

The spaceship construction uses a configuration with four independently operating space engines, which are pivotally attached to the outer side walls of the engine segments 2, 4, 6 and 8. For this purpose, large ring bearings are mounted into the hull wall that have an outer diameter of 154 centimeters. Within their inner opening, a hatch can be inserted for maintenance and repair needs. In combination with correspondingly outside-lying and pressure-resistant engine gondolas or maintenance areas, the engines thus can be reached also from the spaceship inner. To establish such an access connection, the bearing ring of the engine only needs to be pressed pressure-tightly onto the counterpart of the spaceship hull and to be filled with air. Since this air will be lost to space when disconnecting the maintenance connection, such an access should be used only in case of emergencies.

As described, the four engine gondolas can be rotated by electro-mechanical actuators or pneumatic air-cylinders into different positions and can thus exert their thrust in different directions. During a vertical start from the Earth, for example, the thrust vectors of all engine gondolas point vertically downwards whereby the spaceship disk is accelerated upwards in a horizontal orientation. Furthermore, is this engine position also used for the general acceleration or deceleration of the spaceship, since it generates an inertial gravity force inside the spaceship body that acts against the propulsion direction and which is required for the operation of the life support systems.

After the completion of an acceleration phase, the engines serve in addition for the initiation and trimming of the rotation flight. For this, they are at first turned into a horizontal position from which their thrust force can each act in the direction of the desired rotation acceleration. By the afterwards execution of fine engine thrusts, now the rotation movement around the spaceship rotation axis can be initiated or terminated. Furthermore, allow the engines in a rotated flight the compensation of a precession around the spaceship axis via the LCP system, which is why the engine gondolas ideally should be turned back into the acceleration position after a rotation flight has been initiated.

The installation of this concept-specific and very flexible engine suspension can optionally be waived with the use of a conventional chemically driven rocket. Usually, require missions into the near-earth orbit only additional lightweight control thrusters that can be installed also in form of hydrazine-based thrusters directly onto the spaceship hull (see chemical engines in the chapter on energy supply). Moreover, would the choice of a chemical propulsion already today allow the use of the concept spaceship for a long-term flight without having to rely on the specially described high-energy sources and space engines. Only the costs of such a conventionally propulsed mission remain at a very high level due to the used inefficient engines and fuels, since these require the carriage of a large amount of fuel as supporting mass. By this circumstance, the initially required propulsion force of a conventional launching system is – in comparison to the real required propulsion force for the concept spaceship – rapidly increased by a fifty-fold.

Also when using more efficient engines and energy sources, would the costs for such currently available solutions not be substantially lower, as the energy density of high-energy sources is reflected directly in their acquisition price. So could the use of a decay-based energy source although reduce the spaceship mass, but would be the thereby saved launch expenses quickly equalized by the very expensive energy source. However, if such an investment is not avoided, in addition to the permanently available heat energy of a nuclear high-energy source, also further types of propulsion become available for the concept spaceship.

5.2.2 Review of alternative existing Propulsion Solutions

So remains the question: Which propulsion system would be suitable for an entire mission profile of the spaceship? And what energy demand must be covered for this? Starting from a spaceship mass of 23,500 kilograms, in which the assumed weight of a high-energy source and of four space engines have been already taken into account, initially a space engine force of 230,535 Newton is necessary to lift the spaceship against the Earth's gravity. This value already results out of the multiplication of the spaceship weight with the mean acceleration of gravity of 9.81 m/s², which corresponds to the weight force of one kilogram in Newton.

As a first example, the already practically used ion drive should be mentioned here, which operates by the electromagnetic acceleration of ionized xenon gas. Theoretically, such an engine could already be supplied today with the electric energy of a radioisotope generator. But, moreover, the decisive factor for the efficiency of this type of thruster is the available amount of energy with which the ionized gas can be accelerated. If this efficiency is increased, simultaneously also the amount of required xenon gas will be reduced that has to be taken along as supporting mass. A reference efficiency to this can be provided by the dual-stage-4 thruster, which was developed by the European Space Agency ESA and the Australian National University ANU and that already represents a further development stage of the ion thrusters currently in use. The maximum propulsion force of such a DS4G thruster reaches about six Millinewton for which an energy input of 300 watts is required (Bramanti *et al.* 2006). If extrapolating these data to the required space engine force, thus an energy requirement would result of

$$W_{max} = \frac{250\,W \cdot 230{,}535\,N}{0.006\,N}$$

$$W_{max} = 9{,}606\,MW$$

The energy level in the megawatt range seems quite impressive. And in comparison to the performance of earthly energy sources, these dimensions become even more impressive. The hydropower plant of the Hoover-Dam, for example, provides two gigawatts (2,000 megawatts), so that even the entire output of this power plant would not be nearly enough to lift a 23.5-tons spaceship with this thruster type. If one wanted to produce such an amount of energy with the radioisotope generators of Voyager, which were able to generate 470 watts on mission start, nearly 21 million of these RTGs would be required for this. Thus, it is clear that – even if ignoring the disadvantage of a farther necessary supporting mass – solely the required amount of energy disqualifies the ion thruster as capable universal spaceship engine under gravity conditions.

Another possibility for a space propulsion is the use of pure physical energy that puts a focus, in particular, on the avoidance of a supporting mass to be carried along. When leaving therefore the field of chemical-physical drives, the simplest method to propulse a spacecraft mechanically is the usage of a solar sail. In the past, national experiments have already been carried out to test the suitability of a thin plastic or metal sail for this purpose.

Their propulsion principle is based on the force of the solar wind, which provides with its particle radiation pressure an energetically effect comparable to the terrestrial wind energy. Hit these solar particles on a large surface, thus, they consequently exert a slight force in the direction of the solar wind. If the results of the referred experiments are taken, by the use of a solar sail surface of 173.63 square meters, a maximum achievable propulsion force of 1.1 Millinewton can be assumed (Ono *et al.* 2013), but which furthermore is decreased with an increasing distance to the sun. The undeniable advantage of this method lies in the complete independence from a carried fuel. A spaceship using such a drive thus could, in principle, waive all fuel supplies. But within this context, also the disadvantages of this type of propulsion have to be mentioned:

- The achievable propulsion force is extremely low, so that this propulsion form can be used only after an already performed start acceleration.

- The sun sails solely provide a thrust force to move away from the sun. Their usage for a flight into the opposite direction is not possible.

5.2.3 Centrifugal Mass Space Engine

Nevertheless, a space engine with the ability to a low but continuous acceleration would be perfectly suited for the concept spaceship. If, with the available propulsion force, a point would be reached at which the spaceship could just lift off from Earth, already the complete launch from the Earth's surface would be enabled. This can be achieved because of the long intended usage time of the space engines whereby a very slow ascent into space is made possible. With an increasing distance to the Earth, the gravitational effect finally decreases, which is why the speed of the spaceship can gradually further increase. If, for example, a height of 5,000 kilometers has been reached, the Earth's gravity would be already reduced by about 70 percent. Consequently, the available engine energy now can be used for a greater acceleration, which leads – depending on the direction and duration of the thrust – either into an orbit or onto an escape flight course. Moreover, such engine dimensioning generates very smooth acceleration- and deceleration forces in the strength of the Earth's gravity. Because of this, the gravity-based life support systems can be operated during acceleration- and deceleration phases analogous to the landed configuration.

Table 11: Composition of the spaceship mass

Spaceship component	Mass (kg)
Spaceship hull	10,234
Landing gear [a]	50
6 x planting rack á 61.4 kg	369
Thermal insulation (air and kapok fibers)	359
Radiation protection (inner polyethylene shell)	2,236
Meteorite protection (aramid fabric and coating)	129
Nutrient solution (planting racks and tanks)	1,428
Drinking water and electrolyte	66
Living biomass (incl. one person)	1,469
Further components (bioreactors, CINA, living room) [a]	120
4 x space engine á 500 kg [a]	2,000
LiDis I/II minimal:	18,459
Radioisotope generator or cold fusion reactor (optional) [a]	5,000
Water filling of hull cavity (optional)	22,357
LiDis I/II maximal:	45,816

[a] assumed values

If we assume for our future space engines a propulsion type not existing today, the challenges for its construction could not be greater:

- Generation of a total propulsive thrust of 230,535 Newton,

- use of available power from assumed 12,000 watts of electrical or 35,820 watts of thermal energy,

- avoiding the release of a support mass that massively increases the starting weight and that limits the operational usage time.

If we take the calculated available electric power of 12,000 watts, the lifting of a few ton heavy spaceship on Earth would be possible by a motorized rope hoist without any problem. Depending on a selected motor translation, a comparable weight could be lifted even with 10,000 watts within one minute by 2.7 meters. This comparison seems – in association with a spacecraft propulsion – at first unintelligibly. But it shows, which high lifting power would be possible with the given resources, if a much more massive but also spaceship-independent counter mass could be used, which serves to the engines so to say as a free supporting mass.

And in fact, there is already a theoretical spaceship drive that follows such an approach: the space elevator. Briefly circumscribed, such a stationary facility uses a strong and very long cable, which is fixed to the Earth's equator and therefrom rises perpendicularly into the outer space. At its outer end, at a height of more than 35,786 kilometers, a counterweight is located in form of a meteorite or a space station. Since with this distance a geostationary orbit is given, the counterweight can relatively easy be held by a corresponding circumferential speed in a stationary position over the anchoring point on Earth. If its orbit is slightly shifted outwards beyond this, the centrifugal force of the counterweight exceeds the gravitation force of Earth due to its higher circumferential speed, whereby it pulls at its cable outwards and tightens this over the whole length. A space capsule attached to the cable can now easily ascent from the Earth to a geostationary orbit (Edwards 2000).

Here now the chain hoist and the space elevator are following the same principle as both are anchored to an external fix-point and use this as a non along-taken supporting mass. However, the principle of the space elevator uses, instead of a direct repulsion, an outward pulling centrifugal force, which remains by its cleverly positioning always at the same location above the cable anchor.

This is where the theoretical consideration of whether such a rectified centrifugal force can not be generated on a smaller scale begins. For this purpose, a horizontal pendulum with a four degrees forwardly inclined rotation orbit was used as a mechanical trial basis. It had a length of 27.2 centimeters and a mounted weight at its outer end of 220 grams. Its ball-bearing axis was fixed on a wheeled cart, which weighed alone 500 grams and that was standing on a polished granite slab in order to reduce the rolling frictional resistance.

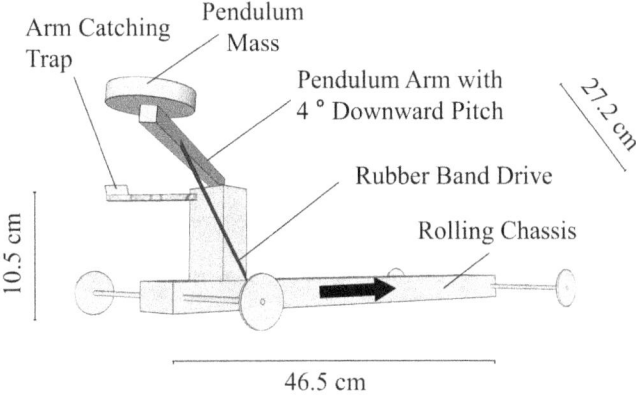

Fig. 38: Side view of a gravitation pendulum

At the beginning of the experiment, the pendulum was located in a raised rest position, which was 135 degrees behind the front pendulum low-point. In order to accelerate the pendulum arm, a weak rubber band was attached at the cart two centimeters laterally of the pendulum axis, which pulled the arm at the begin of the rotation slightly to the front. After the first 45 degrees, the pendulum was accelerated, however, only by the Earth's gravity. When passing through the front position, a main portion of its swing was temporarily transferred into a forward movement of the cart. Then, the pendulum slowed down on the run-out side against the Earth's gravity and stopped at 151 degrees behind the front pendulum low-point.

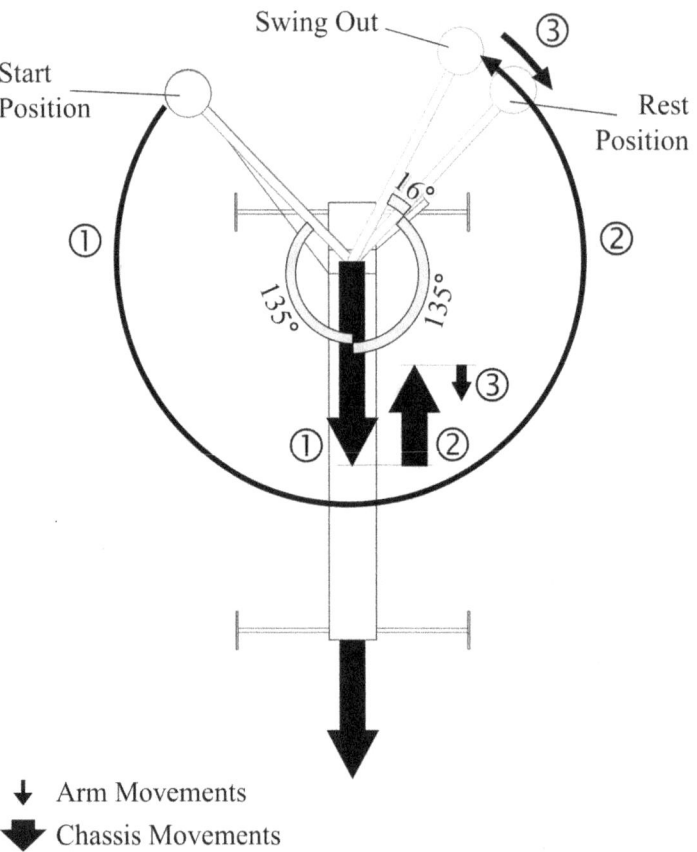

Fig. 39: *Gravitation pendulum endpoints*

Due to the acceleration and deceleration by gravity, such a pendulum motion already converts a part of the horizontal momentum force into a downward-acting vertical force. In the experiment this meant that the forward movement of the cart did not completely stop after the run-out of the pendulum swing, but continued in part as a horizontal propulsion momentum. This effect was further reinforced whilst the pendulum arm hit – after a backwards rotation of 16 degrees – against a catching device whereby an additional pushing pulse was produced in drive direction. This result was achieved also in a second experiment in opposite direction whereby the influence of a possibly existing gradient has been excluded.

If looking at the start- and end position of the pendulum experiment, the propulsion pendulum is after a completed swing at the same vertical height and at the same angle to the longitudinal axis of the cart. If this position is now used as a starting point for a second opposing swing, thus, the forward movement of the cart should be continued in a cyclical movement.

By this means, this technique can thus generate a propulsion force based on the gravity force that acts at 90 degrees to the drive direction. But the described propulsion principle is not inert, which means that to comply with the energy conservation law still forces must be derived to the Earth below the pendulum. Applying this principle now to a free-floating spacecraft, however, the Earth's gravity can no longer be used.

Therefore, here again an artificial gravity is used, whereto the propulsion pendulums described will be mounted inside a rotating centrifuge. This allows with its centrifugal force a derivation of the vertical driving forces against its outer walls, which is comparable to the pulling effect of the counterweight at the end of the space elevator cable. For this purpose, the pendulums will be aligned with their pendulum low point in the direction of the artificial gravity on the opposed centrifuge walls where their propulsion direction consequently acts parallel to the centrifuge axis. While the centrifuge is in rotation, the forward-accelerated pendulums are passing – analogous to the pendulum experiment – simultaneously the fore pendulum position and generate together thus the propulsion force for the spacecraft. A portion of their acceleration and deceleration energy is thereby now redirected over the radially acting centrifugal force, which is why after the completion of each single propulsion swing the spacecraft will be provided with a remaining forward impulse.

Also in this case, repeated cyclic propulsion swings are performed, which maintain the propulsion force continuously. Due to the simultaneous and opposing rotation of the pendulum arms, the torques of their acceleration will be neutralized whereby the centrifuge rotation can take place without unbalance. The drive of this mechanical thrust generation could be done either electrically or mechanically by a steam turbine. Both energy sources would already be available with the presented variants of the high-energy generation.

Such a theoretical propulsion system certainly offers many opportunities for discussion, and its development needs further research. However, it remains to be stated that the today's solutions for purely electric or solar space drives due to their low thrust generation and the required amounts of energy – even for a downsized concept spaceship of the spaceship class described herein (LiDis I) – are not suitable. If the immense costs of the chemical rocket engines are also to be avoided, inevitably a mechanical-physical propulsion form has to be used, which on the one hand side provides a sufficient launch and landing performance and that, in the ideal case, can also be used for a very long operation phase.

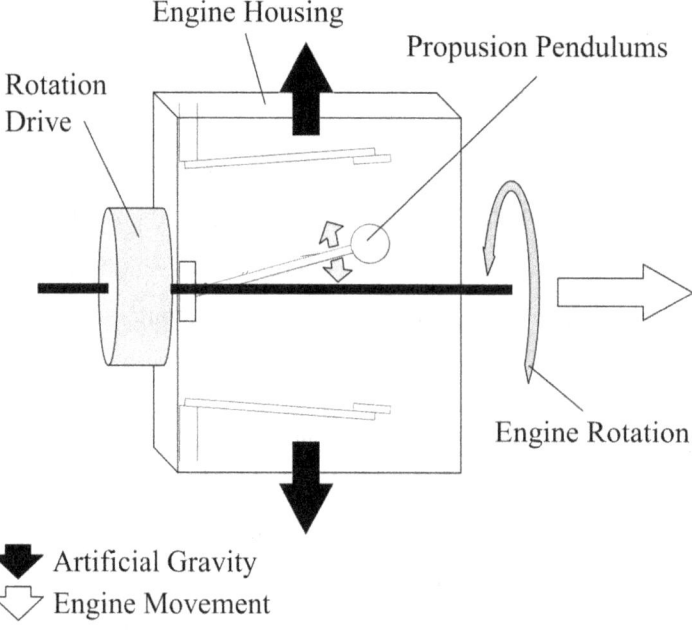

Fig. 40: *Principle of a mass-centrifugal engine*

5.3 Personal Living Space

Adjoining to the described navigation room, segment 1 also comprises a living room for one or two space travelers. Within this outer area, which also serves to accommodate the nutrient solution pump and the air conveyance unit, an area of approximately 3.6 square meters is available from which the base area of the pump has been already deducted. Due to the location of the living room, it can generally be used under full gravity, which, however, may change in its direction with the different flight phases. Therefore, the sleeping site is mounted centrally along the outer wall in form of a pivotable bed that can be tilted in dependency to the gravity direction by 90 degrees. At the wall to segment 2, between the nutrient solution pump housing and the access ladder to the navigation room, the possibility to a water extraction and two small working surfaces are mounted at an angle of 90 degrees to each other. Therefore, these installations can be reached and be used under both possible gravitational directions. The ladder to the navigation room serves during rotation flight phases to a fast reachability of all controls and sensors.

In terms of color, the living room should be different from the sober design of all other spaceship segments, which is why a blue, sky-like ceiling painting and warm wood tones are intended for the wall decoration here.

6 On Board Software

Although 90 percent of all in this concept presented systems are working on a simple mechanical or electrical basis, an additionally in the spaceship integrated electronic can nevertheless be helpful as information storage and control system.

6.1 Electronics Hardware

However, the selection of appropriate hardware for installation in the CINA is subject to certain requirements as well. An anytime available long-term storage, for example, must be insensitive to vibrations, which is why it can only consist of non-mechanical memory cards. A conventional magnetic hard disk thus can be used due to their sensitive mechanics solely as a backup memory for calm flight phases. For the screen display of such a computer system, long-lasting LCD flat screens are currently available. For this reason, the integration of such a display already has been foreseen, which however will be activated only when needed.

Current space missions base their computer systems not exclusively on the latest computer technology. This is because an outdated but already well-proven technology is usually less prone to hardware errors. Furthermore, the coarser structures of these systems are more resistant against the incoming cosmic radiation, which reduces their risk of failure additionally. The in 2011 launched Mars mission of the Curiosity robot probe, for example, was equipped with a RAD750 processor, which was already developed ten years earlier and thus is with its performance capability at about the level of the terrestrial computers of this time (White 2012). In the hope of a very long service life duration, also the computer system of the concept spaceship should be based on two redundant but outdated computer units in which processors of the 80486 command standard from the years 1991 to 1995 were mounted. The developed operating system software is in its current version already designed for this architecture. In addition, another variant is being prepared for the todays increasingly used ARM architecture so that, if applicable, also such computer systems can be integrated. The ARM architecture is very often used in mobile phones or tablets, which is why these processors have already proven to be very robust and energy-saving alternatives.

6.2 Computer Operating System

The software of the computers local environment operator (CLEO) uses for a simplified information storing its own filing structure. For this purpose, the data are stored on the main storage media without fragmentation in order to enable an easier data analysis by eventual finders. In addition, the storage is divided into a word memory and a data memory. A special feature of the CLEO system is its ability to communicate in complete sentences by using written or, in future, also spoken language. By the execution of a continuous communication analysis, it thus can learn new words and is able to store context-related connections.

As described in the section on spaceship navigation, for the verbal interface to CLEO, each a microphone and a speaker are to be installed within the outer and inner ring corridor of each segment, so that in total 16 speakers and microphones must be connected via appropriate amplifier modules. In this way, many information and functions can be requested also without using the screen and the keyboard. In addition, the CINA provides in the navigation room a graphical user interface in which all menus and access pages are displayed according to the single spaceship components. Their display is explicitly not window-based in order, on the one hand, not to overfill the smaller display and, on the other hand, to enable a nested information display of the individually addressed topics. This means for the communication with CLEO that there will be always one topic in the communicative focus, out of which – similar to a continuous narrative – automatically a path to further programs and sub-programs will be followed that each represent their own specific data. In combination with a parallel implemented entity communication, which is based on thematic and logical cross-references, thus an intuitively controlled and highly efficient user interaction is created.

Table 12: Energy demands of the CLEO system

Computer component	Power demand (W)
Computer unit (ARM technology)	3.2
Microphone amplifier	0.1
Speaker amplifier	
Sensor- and switch interface	or 30.0
LCD screen	3.5
External hard drive for backup	5.6
Total:	33.3
Radio communication unit for data exchange at maximum	282.9
Communication energy [a]:	282.9

[a] Plant illumination has to be deactivated in parallel

The CLEO system executes several independent tasks in the background for a continuous analysis of the incoming sensor data and stores these in historicized form on the long-term storage. In dependency thereto, automated actions subsequently can be triggered via the output interfaces as well. The kernel underlying this process has already been programmed to a large extent. Therein, the drivers required for the hardware access can be developed and tested simultaneously with the spaceship systems in order to adapt their functions precisely to the conditions. For this purpose, the system offers an implemented object and variable administration, which is easily accessible for all software components via a dialect of the BASIC programming language. The kernel and the integrated drivers, on the other hand, were machine-oriented implemented by using the Assembler programming language. All already completed source codes are kept very simple due to the avoidance of an external operating system. This is a big advantage in the case of necessary corrections or further developments, since the software in its complexity can still be analyzed and understood by a single person.

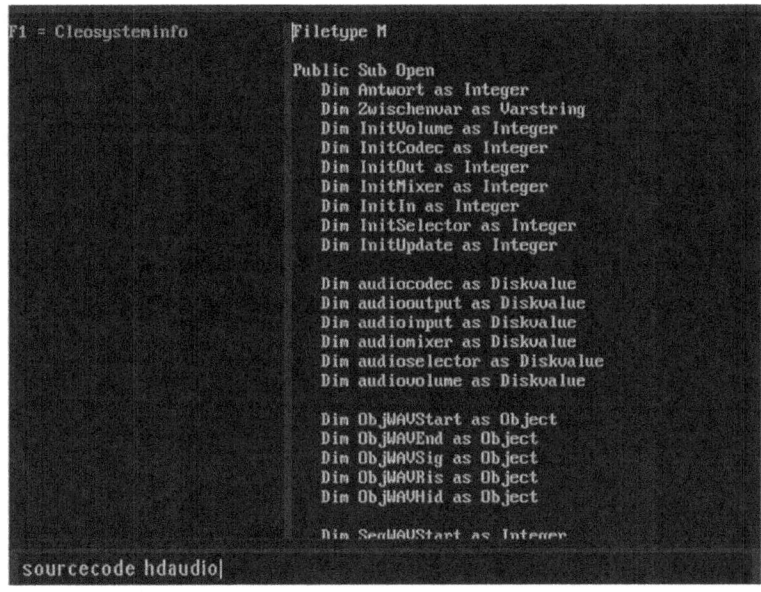

Fig. 41: Screenshot of a CLEO application

The basic hardware requirements of the CLEO operating system consist for the version of the 80486 architecture essentially of:

- 33 MHz processor. Processors with more performance may be used as well.
- 256 MB RAM, of which 1 MB is used by the kernel and the hardware interfaces.
- 1-128 GB memory card respectively hard drive space, of which about 1.7 MB are used by the operating system.

In order to structure its internal organization, the CLEO system uses a hardware-assisted memory mapping and, if available, the protected mode of the used processor. For processor architectures without a protected mode, corresponding system management functions will be fully simulated by the software. This results in the use of standardized memory segments as well as the ability to multitask by using the round robin proceed.

Another special feature of this operating system is the embedding of a "logical conclusion", whose permanent logic analysis can further develop a focused thought by its own. Similar to a person's pathway of thinking, CLEO draws conclusions from an initial statement by bringing its content in the context of her own stored connections and finishes this thought cycle mostly with a new statement. This she can put in the following again into question whereby she finds more and more new thinking approaches. In addition to this, there are also mechanisms integrated that upon reaching certain thresholds in terms of importance, novelty or CLEOs feeling level, can also trigger independent actions such as an information output or a theme-related question.

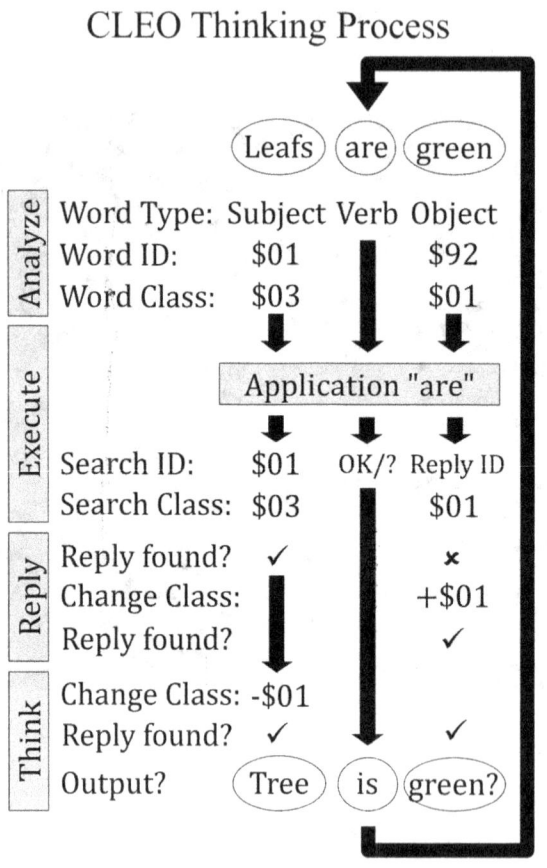

Fig. 42: CLEO thinking process

The contextual and assessing parameterization of properties is automatically applied to all entries of the main storage medium. As a result, the data or information stored in the form of words can have logical relationships to each other and can also influence the mentioned thresholds to positive or negative. The Tables 13a, 13b, and 13c show exemplarily the threshold variables of the CLEO kernel, the property parameters of a word entry as well as an example of a logical connection between single words.

With the current development state of the system, also some complementary additional components have been introduced. A system-based program compiler, for example, serves to the machine-based implementation of all included source codes. With this tool,

CLEO can be already today further developed with her own applications. Additional software modules in the form of video- and USB drivers, text- and source code editors, as well as the first version of a menu system round off the currently available components. The main storage medium, moreover, has been pre-filled in this version with 274 words in order to enable a first sentence-based input and output. At the moment, this basic vocabulary consists of German and English contents. A future language module can also use this word-based and language-based filing system directly, since its function decomposes the spoken language into phonemes, which can be stored therein as character-coded words in their own language variant.

A complete image of the CLEO operating system, including all source code files and drivers, currently requires an own storage volume of 1.7 megabytes. For testing purposes, this can be provided by the author on a full drive image. A variant of the CLEO compiler, which converts the given source code also directly into ARM machine code, is thereto in preparation as well.

With the given functions and the provided sensor data, even CLEO can continue the thought of researching and discovering. She becomes so – far away from home – in the same sense an ambassadress of humankind.

Table 13a: Threshold variables of the CLEO kernel

Offset	Variable name
Byte $000	CLEO-Status
	Bit 0 CLEO-Mode
	Bit 7-1 Number of nested applications
Byte $003	Language key
Byte $005	Emotional desire parameter
Byte $006	Question desire parameter
Byte $007	Statement desire parameter
Byte $008	Reply desire parameter
Byte $009	Execution desire parameter
Byte $00a	Parameter recognized subject
Byte $00b	Parameter recognized object
Byte $00c	Parameter recognized verb
Byte $00d	Reply code after execution
Byte $28a	Word class 1 searched word
Byte $28b	Word class 2 searched word
Byte $28c	Word class 3 searched word
Byte $28d	Word class 4 searched word
Byte $38f	Bit 0-1: Current search level
	Bit 2-3: Searched class direction
	(%00 no search
	%01 higher level
	%10 lower level
	%11 same level)
	Bit 4-5: Counter of positive searches
	Bit 6 : New word found
	Bit 7 : AI activated

Table 13b: Example parameter of a logic word

Offset	Parameter name
Byte $00-$0f	Entry name (max. 14 chars)
Byte $11	Language key ($00 = neutral)
Byte $12	Word class 1 ($00 = neutral)
Byte $13	Word class 2 (object parameter)
Byte $14	Word class 3 (driver number)
Byte $15	Word class 4
Byte $16	Word type training 1
	Bit 8-7: Focus on key parameter
	Bit 6: Word type flag
	Bit 5: Language
	Bit 4: Emotional desire
	Bit 3: Question desire
	Bit 2: Reply desire
	Bit 1: Execution desire
Byte $17	Word type training 2
	Bit 8: Training gender
	Bit 7: Training word type
	Bit 5-4: Gender flag
	Bit 3-1: Wort type
Word $18	Function parameter for application
Byte $1a	Bit 8-7: 0
	Bit 6-1: Influence on emotional desire
Byte $1b	Influence on question desire
Byte $1c	Influence on statement desire
Byte $1d	Influence on reply desire
Byte $1e	Influence on execution desire
Byte $1f	Word description
	Bit 8-6: Word lenght / 2
	Bit 5-4: Gender flag
	Standard entry
	%00 Object (what?)
	%01 Female (who?)
	%10 Male (who?)
	%11 Situation (how?)
	Verb entry
	%00 Word for "is"
	%01 Assignment
	%10 Execution / Change
	%11 Provide / Request
	Question word
	%00 Object (what?)
	%01 Person (who?)
	%10 Choice (which?)
	%11 Count (how many?)
	Attribute
	%00 undefined article
	%01 Female article
	%10 Male article
	%11 Conjunction
	Bit 3-1: Wort type
	$00 Unknown
	$01 Verb
	$02 Substantive
	$03 Adjective
	$04 Attribute
	$05 Question word
	$06 Word for disagreement
	$07 Word for availability

Table 13c: Logic connections of single logic table entries

All stored words are classified by four logic classes (word parameter $12-$15). The first class represents thereby the highest and the fourth class the lowest level of classification.

Class 1	Class 2	Class 3	Class 4
$00 Unbound	$00 Unbound	$00 Unbound	$00 Unbound
(Class 1: $01-$20 Function)			
$01 deinstall			
$03 search			
$04 read			
$05 write			
$06 delete			
$07 increase			
$08 decrease			
(Class 1: $21-$7f Basis classes)			
$31 Nature			
	$c0 Tree		
		$01 Leaf	
			$01 Chlorophyll
$32 Earth			
$33 Physics			
$34 Chemistry			
$35 Astronomy			
	$01 Sun		
		$01 Planet	
			$01 Moon
$36 Humans			
	$c1 Language		
$41 Technic			
$42 Navigation			
$43 Manipulation			
$44 Vision			
(Class 1: $80-$ff Output drivers)			
$82 Screen 1			
$85 Sensor interface			
$86 Switch interface			
$89 Audio output			
	$91 Volume control		
	$94 Speakers		
	$95 Headphones		
	$9a Line-Out		
	$98 Microphone		
	$99 Headset		
	$9b Line-In		
$c1 Clock			
	$3a Time		
	$2e Date (German format)		
	$2f Date (English format)		

7 Conclusions

With the development of this concept, I tried to introduce a new approach for a sustainable spaceship construction. The focus on this work certainly is more related to the spaceship design and the useful combination of life support components via which the human needs can be largely supported. Furthermore, the described solution offers an ideal integration of necessary protection shieldings, space engines, and a closed biospheric life support system.

With the given design stage, most of the described components require a further development in order to achieve an applicable version for practical use. Since this concept fundamentally changes the current kind of spaceship design, the foreseen construction and its related component-composition have not yet been realized in this form. Therefore, I put even more importance on the plausibility by using the cited publications and by verifying my own assumptions, if this was possible and necessary, with own experiments.

Summarizing the results of the presented design and technology research, the construction, launch, and operation of a fully sustainable and artificial gravity supporting spaceship is possible if the assumptions for the described components are met and the necessary space environment conditions can be provided.

In view of the increasing demand for more sustainable spacecraft, I expect that in the future numerous further research papers will have the same focus.

Acknowledgements

Over the past ten years that were used to create this concept, several persons, but also developments in our world and technology, inspired me in different ways. Now, looking backwards on this time, there are even more some special persons that were outstanding with their understanding and their backing to me.

I would like to thank my friends, who supported me in some business related project activities and always accepted nothing less than excellence from me.

Furthermore, I must express my very profound gratitude to my family and to my wife for providing me with unfailing support and continuous encouragement throughout my years of researching and through the process of writing this concept. This accomplishment would not have been possible without them. Thank you.

Bibliography

Alves, H. J. et al. 2013, Overview of hydrogen production technologies from biogas and the applications in fuel cells, *International journal of hydrogen energy* 38(13), 5,215-5,225.

Azcón-Bieto, J. 1983a, Inhibition of photosynthesis by carbohydrates in wheat leaves, *Plant physiology* 73(3), 681-686.

Azcón-Bieto, J., Lambers, H., Day, D. A. 1983b, Effect of photosynthesis and carbohydrate status on respiratory rates and the involvement of the alternative pathway in leaf respiration, *Plant physiology* 72(3), 598-603.

Barber, T., Cowley, R. 2002, Initial Cassini propulsion system in-flight characterization, *38th AIAA/ASME/SAE/ASEE Joint Propulsion Conference & Exhibit*.

Barta, D. J. et al. 1992, Evaluation of light emitting diode characteristics for a space-based plant irradiation source, *Advances in Space Research* 12(5), 141-149.

Bennett, G. 2006, Space nuclear power: opening the final frontier, *4th International Energy Conversion Engineering Conference and Exhibit (IECEC)*.

Bernal, J. D., Maurois, A., Radhakrishnan, S. 1929, The world, the flesh and the devil, *Kegan Paul, Trench, Trubner*, 8-11.

Birur, G. C. et al. 2001, Spacecraft Thermal Control, *Encyclopedia of Physical Science and Technology*, Third Edition (as of March 30, 2001, https://ntrs.nasa.gov/archive/nasa/casi.ntrs.nasa.gov/20010091676.pdf, downloaded at May 3, 2017), 6-8.

Blanchard, A. et al. 1999, Updated Critical Mass Estimates for Plutonium-238, *Savannah River Site*, http://sti.srs.gov/fulltext/ms9900313/ms9900313.html (downloaded at June 16, 2017).

Bloom, A. J., Sukrapanna, S. S., Warner, R. L. 1992, Root respiration associated with ammonium and nitrate absorption and assimilation by barley, *Plant Physiology* 99(4), 1,294-1,301.

Bouallagui, H. et al. 2004, Effect of temperature on the performance of an anaerobic tubular reactor treating fruit and vegetable waste, *Process Biochemistry* 39(12), 2,143-2,148.

Bramanti, C. et al. 2006, The innovative dual-stage 4-grid ion thruster concept–theory and first experimental results, *Fifty-seventh International Astronautical Congress, number* IAC-06-C4.

Butikov, E. 2006, Precession and nutation of a gyroscope, *European journal of physics* 27(5), 1,071.

Carrillo, J. G. et al. 2012, Ballistic performance of thermoplastic composite laminates made from aramid woven fabric and polypropylene matrix, *Polymer Testing* 31(4), 512-519.

Carvajal, M. 2010, Investigation into CO2 absorption of the most representative agricultural crops of the region of murcia, *CSIC (Consejo Superior de Investigaciones Cientificas)*, Madrid, Spain.

De la Fuente, H. et al. 2000, TransHab-NASA's large-scale inflatable spacecraft, *41st Structures, Structural Dynamics, and Materials Conference and Exhibit* (as of April 6, 2000, https://ntrs.nasa.gov/archive/nasa/casi.ntrs.nasa.gov/20100042636.pdf, downloaded at May 4, 2017), 2-5.

Edwards, B. C. 2000, Design and deployment of a space elevator, *Acta Astronautica* 47(10), 735-744.

El-Genk, M. S., Tournier, J.-M. P. 2004, "SAIRS"—Scalable Amtec Integrated Reactor space power System, *Progress in Nuclear Energy* 45(1), 25-69.

Engrand, C. 2001, Meteorites and cosmic dust: Interstellar heritage and nebular processes in the early solar system, *EPJ Web of Conferences. EDP Sciences, 2011*, 05001.

Engstrom, R. W. 1980, Photomultiplier handbook, *RCA Corp*.

Fischer, H. 2016, Potenzial von Halophyten für stoffliche und energetische Nutzung am Beispiel von Salicornia, *Universität Bonn*.

Glenn, E. P. et al. 1992, Climate: growing halophytes to remove carbon from the atmosphere, *Environment: Science and Policy for Sustainable Development* 34(3), 40-43.

Grammier, R. S. 2009, A look inside the Juno Mission to Jupiter, *Aerospace conference*, 2009 IEEE. IEEE, 1-10.

Hamacher, T., Bradshaw, A. M. 2001, Fusion as a future power source: recent achievements and prospects, *18th world energy congress*.

Havenith, G., Holmér, I., Parsons, K. 2002, Personal factors in thermal comfort assessment: clothing properties and metabolic heat production, *Energy and buildings* 34(6), 581-591.

Henry, C. J. K. 2005, Basal metabolic rate studies in humans: measurement and development of new equations, *Public health nutrition* 8(7a), 1,133-1,152.

Herbst, E. 2001, The chemistry of interstellar space, *Chemical Society Reviews* 30(3), 168-176.

Holderman, M., Henderson, E. 2011, Nautilus-X Multi-Mission Space Exploration Vehicle, *NASA Johnson Space Centre*, Concept Presentation, 1-28.

Kim, J. et al. 1995, Modeling of proton exchange membrane fuel cell performance with an empirical equation, *Journal of the Electrochemical Society* 142(8), 2,670-2,674.

Linke, B. et al. 2006, Grundlagen und Verfahren der Biogasgewinnung, *Leitfaden Biogas*, 13-25.

McAlister, D. R. 2012, Gamma ray attenuation properties of common shielding materials, *University Lane Lisle*, USA.

Miller, J. et al. 2003, Benchmark studies of the effectiveness of structural and internal materials as radiation shielding for the international space station, *Radiation research* 159(3), 381-390.

Miramonti, L. 2010, Solar neutrinos: from their production to their detection, PoS, 030.

Monteiro, S. N. et al. 2012, Thermogravimetric stability of polymer composites reinforced with less common lignocellulosic fibers–an Overview, *Journal of Materials Research and Technology* 1(2), 117-126.

Mund, K., Richter, G., von Sturm, F. 1977, Titanium-Containing Raney Nickel Catalyst for Hydrogen Electrodes in Alkaline Fuel Cell Systems, *Journal of the Electrochemical Society* 124(1), 1-6.

Munns, R., James, R. A. 2003, Screening methods for salinity tolerance: a case study with tetraploid wheat, *Plant and soil* 253(1), 201-218.

NASA JPL 2015, Voyager Mission Operations Status Report, as of January 2015, https://voyager.jpl.nasa.gov/mission/weekly-reports/index.htm (downloaded at June 14, 2017).

Nelson, M. et al. 1994, Atmospheric dynamics and bioregenerative technologies in a soil-based ecological life support system: initial results from Biosphere 2, *Advances in Space Research* 14(11), 422-423.

Nelson, M. et al. 2010, Closed ecological systems, space life support and biospherics, *Environmental Biotechnology*, Humana Press, 517-565.

Novikova, N. D. 2004, Review of the knowledge of microbial contamination of the Russian manned spacecraft, *Microbial ecology* 47(2), 127-132.

Ono, G. et al. 2013, Development of Mission Devices and Sub-systems on Sail for World's First Solar Power Sail IKAROS, *The Journal of Space Technology and Science* 27(1), 38-53.

Osmond, C. B. 1987, Photosynthesis and carbon economy of plants, *New Phytologist* 106(s1), 161-175.

Pustovalov, A. 2007, Role and prospects of application of RTG on base of plutonium-238 for planetary exploration, *5th European Conference on Thermoelectrics*, Odessa, Ukraine.

Rajendrudu, G., Prasad, J. SR, Das, VS R. 1986, C3-C4 Intermediate Species in Alternanthera (Amaranthaceae) Leaf Anatomy, CO2 Compensation Point, Net CO2 Exchange and Activities of Photosynthetic Enzymes, *Plant Physiology* 80(2), 409-414.

Rayman, M. D. et al. 2006, Dawn: A mission in development for exploration of main belt asteroids Vesta and Ceres, *Acta Astronautica* 58(11), 605-616.

Romanishin, W. 2006, An Introduction to Astronomical Photometry Using CCDs, *University of Oklahoma* 17.

Rroço, E., Mengel, K. 2000, Nitrogen losses from entire plants of spring wheat (Triticum aestivum) from tillering to maturation, *European Journal of Agronomy* 13(2), 101-110.

Sakabe, S. et al. 2004, Generation of high-energy protons from the Coulomb explosion of hydrogen clusters by intense femtosecond laser pulses, *Physical Review* A 69, 023203.

Schunk Bahn- und Industrietechnik GmbH 2007, Handbuch für Schunk Brennstoffzellen-Stacks, as of November 2007, https://www.fh-bielefeld.de/multimedia/Fachbereiche/Ingenieurwissenschaften+und+Mathematik/Dokumente/Labore/Energiewirtschaft+und+Regenerative+Energien/Handbuch_FC42_Stack-p-49712.pdf (downloaded at June 12, 2017), 33-36.

Sheehan, D. P. 2011, Preface: Second law of thermodynamics: Status and challenges, *AIP conference proceedings,* AIP, 2011.

Speight, J. G. et al. 2005, Lange's handbook of chemistry, *New York: McGraw-Hill,* 2005.

Stewart, W. M. et al. 2005, The contribution of commercial fertilizer nutrients to food production, *Agronomy Journal* 97(1), 1-6.

Stone, E. C. et al. 2005, Voyager 1 explores the termination shock region and the heliosheath beyond, *Science* 309(5743), 2,017-2,020.

Toivola, M., Napari, I., Vehkamäki, H. 2009, Structure of water–sulfuric acid clusters from molecular dynamics simulations, *Boreal Environ. Res* 14, 654-661.

Tripathy, B. C. et al. 1996, Growth and photosynthetic responses of wheat plants grown in space, *Plant physiology* 110(3), 801-806.

Vlek, P. L. G., Stumpe, J. M., Byrnes, B. H. 1980, Urease activity and inhibition in flooded soil systems, *Nutrient Cycling in Agroecosystems* 1(3), 191-202.

Voumbo, M. L. et al. 2010, Characterization of the thermophysical properties of kapok, *Research Journal of Applied Sciences, Engineering and Technology* 2(2), 143-148.

Vymazal, J. 2007, Removal of nutrients in various types of constructed wetlands, *Science of the total environment* 380(1), 48-65.

Wang, X. et al. 2012, Optimizing feeding composition and carbon–nitrogen ratios for improved methane yield during anaerobic co-digestion of dairy, chicken manure and wheat straw, *Bioresource technology* 120, 78-83.

White, M. 2012, Scaled CMOS reliability and considerations for spacecraft systems: bottom-up and top-down perspectives. *Reliability Physics Symposium (IRPS),* 2012 IEEE International. IEEE, 4B. 4.1-4B. 4.5.

Zeitlin, C. et al. 2013, Measurements of energetic particle radiation in transit to Mars on the Mars Science Laboratory, *Science* 340(6136), 1,080-1,084.

www.ingramcontent.com/pod-product-compliance
Lightning Source LLC
Chambersburg PA
CBHW081815220526
45470CB00007B/2322